Industrial Railways and Locomotives

of

South Yorkshire

The Coal Industry 1947-1994

INDUSTRIAL RAILWAY SOCIETY

Published by the **INDUSTRIAL RAILWAY SOCIETY**
at 24, Dulverton Road, Melton Mowbray, Leicestershire, LE13 0SF

www.irsociety.co.uk

ISBN 1 901556 43 3

© Industrial Railway Society 2007

COVER PHOTOGRAPH

N2 (HC 1857 of 1952) at North Gawber Colliery on 5th December 1973   (Andrew Smith)

Printed by Print Rite, 31 Parklands, Witney, Oxfordshire, OX29 8HX

CONTENTS

Introduction

Maps

INTRODUCTION

The origin of this book lies in part of the text of the "Pocket Book No. 8 Industrial Locomotives of Yorkshire (East & West Riding)" published in 1955. It is always our intention to publish full historical Handbooks covering all the United Kingdom and this is progressing on a county by county basis. In order to make progress, the Committee sanctioned the production of Interim Pocket Books (IPB) for areas where research undertaken had been limited and where publication of a full historical Handbook was many years away. This project evolved and the title has been changed to Preliminary Handbook.

As there is no possibility of a full Handbook covering all industrial railways and locomotives in South Yorkshire being produced at this time this volume is the NCB section that would appear in that Handbook together with maps and a representative selection of Iphotographs.

A list of common abbreviations are also available on our web site (www.irsociety.co.uk).

We ask the following of those who purchase or borrow a copy of this publication:

Please respect our copyright in this publication. Whilst much of the information in this book is in the public domain, some has come from the private records of members and individuals on the understanding that it is used for the benefit of the Society and other r non profit making bodies.

ACKNOWLEDGMENTS

The publication of this book was made possible thanks to the help of all members of the Society's production team. Over many years staff and management of all the locomotive building companies and many of the collieries have given generously of their time to assist with our many enquires, their contribution has made this volume all the more accurate. We extend our gratitude and a very big thank you to them. The maps were drawn up by Roger Hateley.

We have taken reasonable steps to verify the accuracy of the information in this book but without doubt it will contain errors or omissions. Any information that you can supply that can add to, amplify, or correct any of the details contained in this Handbook is most welcome. Please do so by notifying:

Mr I.R. Bendall
46 Orson Drive
Wigston
Leicestershire
LE18 2EJ

Or email: records@irsociety.co.uk

BRITISH COAL CORPORATION - NATIONAL COAL BOARD
Includes Ministry of Fuel & Power, Directorate of Opencast Coal Production.

On the nationalisation of the coal industry on 1st January 1947, the assets of the coal companies were vested in the National Coal Board. Small mines with less than 30 men underground could be excluded and operate under NCB license. Those companies nationalised had the option to include or exclude ancillary operations such as Brickworks. Those in South Yorkshire were generally included. The title 'National Coal Board' remained in use until 1986, when the trading name 'British Coal' was adopted. The legal title became British Coal Corporation from 5th March 1987.

DEEP MINES ORGANISATION
Collieries absorbed by the NCB on vesting day were placed in an organisation comprising eight Divisions, each of which was divided into a number of Areas. Each area was in turn sub-divided into Groups. However, Groups had little railway significance and reorganisations were frequent, hence these early Groups are disregarded in IRS publications.

Working collieries in what became the county of South Yorkshire were all included in the North-Eastern Division. They initially formed part of Areas No.1, No.4 and No.6 and comprised the whole of Areas No.2, No.3 and No.5. Holbrook (closed) Colliery was located in Derbyshire throughout its existence, being part of East Midlands Division, Area No.1 in its short NCB life. However, its site later became part of the County of South Yorkshire and a listing is therefore included in this volume for completeness.

The first of a number of subsequent reorganisations became effective from 26th April 1964 when Area No.4 was abolished and its South Yorkshire collieries reallocated to Area No.5 and its Central Workshops to Area No.6. North Eastern Division was renamed Yorkshire Division c1965. The next change took effect from 26th March 1967 when a major national reorganisation replaced Divisions by eighteen 'New style' Areas. Surviving South Yorkshire collieries were placed in one or other of the Doncaster, Barnsley or South Yorkshire Areas. The first two of these also included collieries in what became West Yorkshire and the last collieries in Nottinghamshire.

From 1st October 1985, Doncaster and Barnsley areas were disbanded and their collieries became part of South and North Yorkshire Areas respectively. The next reorganisation, effective from 1st April 1990, drastically reduced Area staff and much day to day control reverted to collieries. Areas became known as Groups but South Yorkshire was otherwise unaffected. As contraction of the industry continued, the next change saw North Yorkshire Group abolished from 1st October 1991. Its collieries were divided between Selby and South Yorkshire Groups.

On 1st September 1993, those collieries in South Yorkshire that were assessed to have a future became part of a new Northern Group, with the remainder passing to a Closing Collieries Group.

At Vesting Day many collieries in South Yorkshire included the word MAIN in their titles. Over the years this was progressively dropped in most cases by the NCB except where it avoided confusion with collieries of the same name in other counties. Markham and Houghton are examples. In this publication MAIN is used sparingly and usually for the same reason, or where the colliery closed before the name changed.

WORKSHOPS ORGANISATION
Area Central Workshops were part of the deep mines organisation until June 1967, after which they passed to the direct control of National Headquarters.

COKING, TAR AND MANUFACTURED FUEL PLANTS
Administered by the Divisions of the Deep Mines organisation until 1st January 1963 when they were combined in a new Coal Products Division. From 1st April 1973 this operated through wholly owned subsidiaries, one of which, National Smokeless Fuels Ltd, took over coking and fuel plants. With the change from National Coal Board to British Coal Corporation from 5th March 1987, they passed to a Coal Products Group. A number of coking plants in South Yorkshire which were on, or adjacent to, colliery premises were not vested in the NCB and remained in private hands until closure.

BRICKWORKS
Administered by the Divisions of the Deep Mines organisation until 1st January 1962 when they were combined nationally, to form the Brickworks Executive. This operated through subsidiary companies, the one covering South Yorkshire being known as the Midland Brick Co Ltd. These companies were sold to the private sector on 25th November 1973. As with coking plants, some works on colliery premises were not vested.

OPENCAST SITES AND DISPOSAL POINTS
The first large scale opencast coal workings date from the war years and were operated from 1942 by the Ministry of Fuel & Power (and for a period, the Ministry of Works) through its Directorate of Opencast Coal Production, who provided the management function, with Public Works contractors working the sites and disposal points. The NCB set up its Opencast Executive which took over these activities from 1st April 1952 and operated in exactly the same way. Hence it is convenient to cover the short pre-NCB history of these sites in this section. The Opencast Executive was sub divided into Regions That covering Yorkshire was first designated North Eastern then Central and finally Central East Region. With the exception of an early site at Wentworth, rail traction has not been used on the working sites but was confined to Disposal Points where coal was screened and loaded for sale. Locomotives have either been the property of the NCB or of the contractor operating the Disposal Points. South Yorkshire locomotives were a mixture of the two.

PRIVATISATION
During the 1990s, the industry continued to decline with many collieries being closed and others placed on care and maintenance. The collieries and opencast operations throughout the UK that remained in production were offered for sale. Of those collieries on care an maintenance, most subsequently closed, but others in South Yorkshire were purchased either by RJB Mining (UK) Ltd or by Coal Investments Ltd, passing into the new ownerships on 30/12/1994, and resumed production.

LOCATION LISTINGS
Locations which have made any significant use of rail track, surface or underground, standard or narrow gauge, are listed below alphabetically, together with brief notes of the rail system and full details of locomotives where appropriate. The following abbreviations are used to save space when showing in which part of the organisation a location was placed at any given period:-

    EM1   East Midlands Division, Area No.1 (Chesterfield).
    NE1   North Eastern Division, Area No.1 (Worksop).
    NE2   North Eastern Division, Area No.2 (Doncaster).
    NE3   North Eastern Division, Area No.3 (Rotherham).
    NE4   North Eastern Division, Area No.4 (Carlton).
    NE5   North Eastern Division, Area No.5 (South Barnsley).
    NE6   North Eastern Division, Area No.6 (North Barnsley).
    SM&MDU   North Eastern Division, Small Mines & Mines Drainage Unit.
    NYK   North Yorkshire Area.
    DCR   Doncaster Area.
    BNY   Barnsley Area.
    SYK   South Yorkshire Area.
    NYG   North Yorkshire Group.
    SYG   South Yorkshire Group.
    SEL   Selby Group.
    NG   Northern Group.
    CCG   Closing Collieries Group.
    HQ   National Headquarters, Workshops Organisation.
    CPD   Coal Products Division (later Group).
    OE   Opencast Executive.
    MFP   Ministry of Fuel & Power, Directorate of Opencast Coal Production.

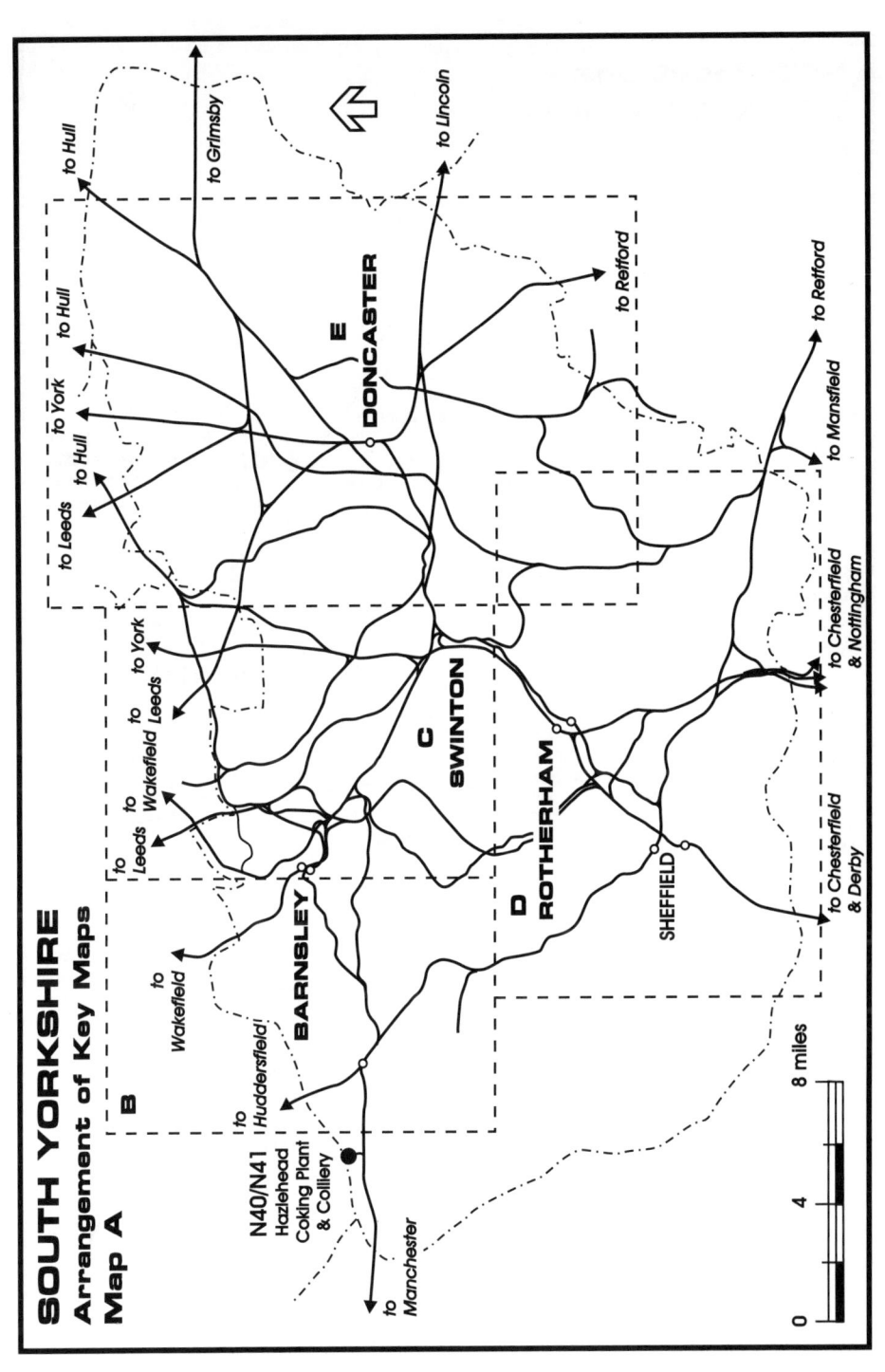

SOUTH YORKSHIRE
Arrangement of Key Maps
Map A

B

N40/N41
Hazlehead
Coking Plant
& Colliery

to
Huddersfield

BARNSLEY

to
Wakefield

to
Manchester

C

SWINTON

to
Leeds

to
Wakefield

to
Leeds

to York

D

ROTHERHAM

SHEFFIELD

to Chesterfield
& Derby

to Chesterfield
& Nottingham

to Mansfield

to Retford

E

DONCASTER

to Leeds

to Hull

to York

to Hull

to Hull

to Grimsby

to Lincoln

to Retford

to Retford

0   4   8 miles

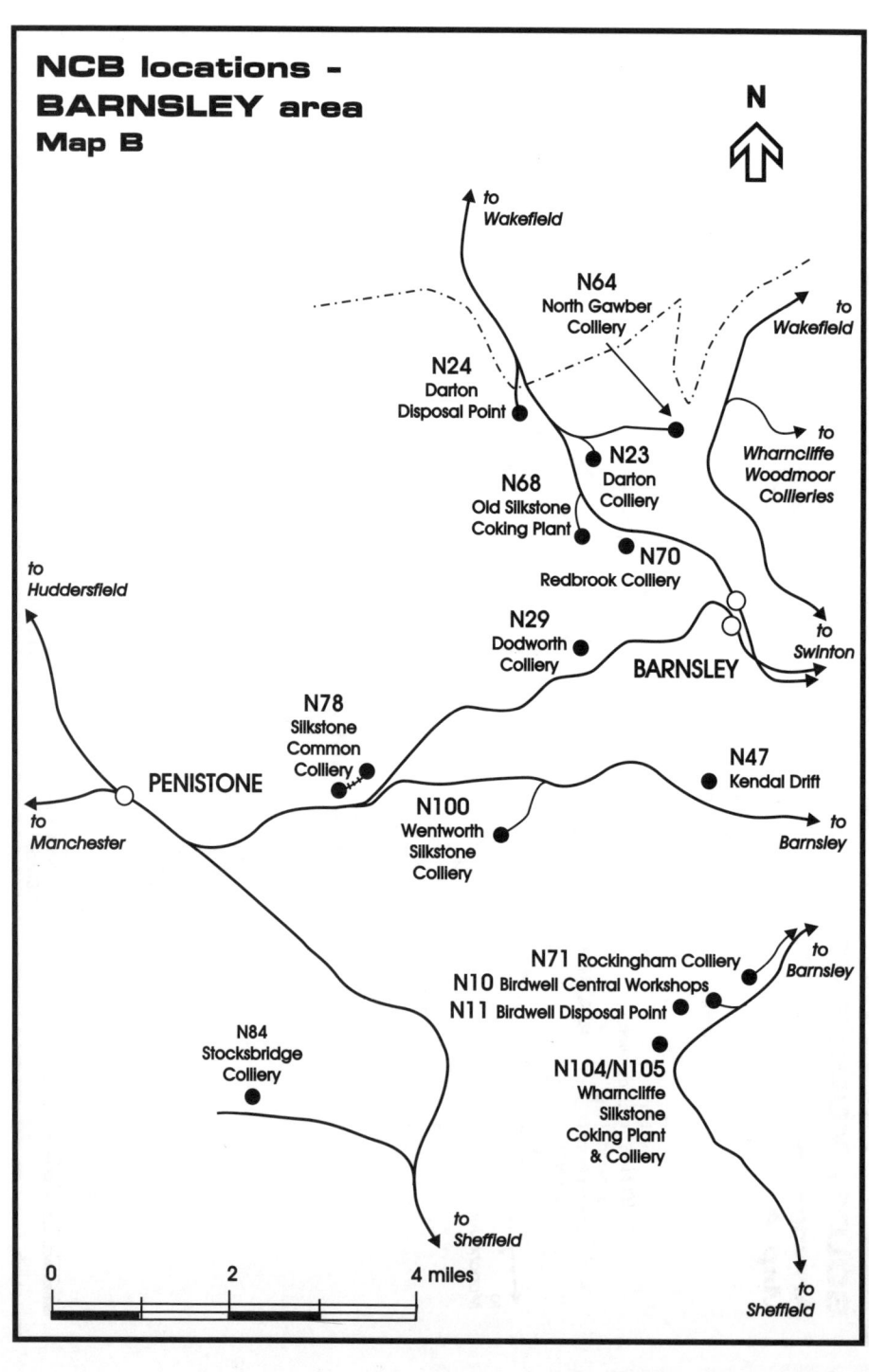

NCB locations –
BARNSLEY area
Map B

N

to Wakefield

N64
North Gawber
Colliery

to Wakefield

N24
Darton
Disposal Point

to
Wharncliffe
Woodmoor
Collieries

N23
Darton
Colliery

N68
Old Silkstone
Coking Plant

N70
Redbrook Colliery

to Huddersfield

N29
Dodworth
Colliery

BARNSLEY

to
Swinton

N78
Silkstone
Common
Colliery

PENISTONE

N47
Kendal Drift

N100
Wentworth
Silkstone
Colliery

to
Manchester

to
Barnsley

N71 Rockingham Colliery
N10 Birdwell Central Workshops
N11 Birdwell Disposal Point

to
Barnsley

N84
Stocksbridge
Colliery

N104/N105
Wharncliffe
Silkstone
Coking Plant
& Colliery

to Sheffield

0   2   4 miles

to
Sheffield

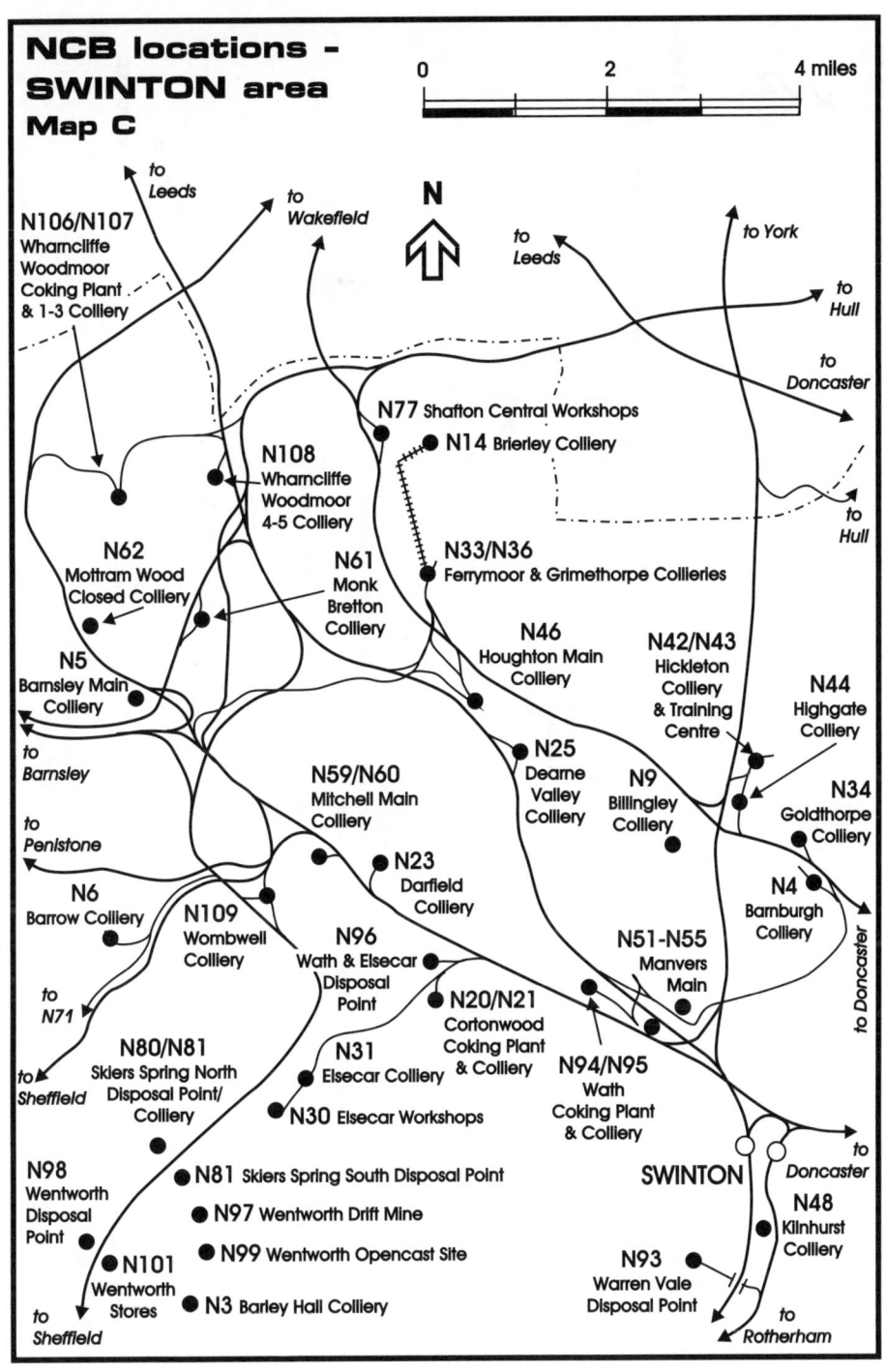

NCB locations – SWINTON area Map C

0 2 4 miles

N

to Leeds
to Wakefield
to Leeds
to York
to Hull
to Doncaster
to Hull

N106/N107 Wharncliffe Woodmoor Coking Plant & 1-3 Colliery

N77 Shafton Central Workshops

N14 Brierley Colliery

N108 Wharncliffe Woodmoor 4-5 Colliery

N62 Mottram Wood Closed Colliery

N61 Monk Bretton Colliery

N33/N36 Ferrymoor & Grimethorpe Collieries

N5 Barnsley Main Colliery

to Barnsley

to Penistone

N46 Houghton Main Colliery

N42/N43 Hickleton Colliery & Training Centre

N44 Highgate Colliery

N25 Dearne Valley Colliery

N9 Billingley Colliery

N34 Goldthorpe Colliery

N59/N60 Mitchell Main Colliery

N23 Darfield Colliery

N6 Barrow Colliery

N109 Wombwell Colliery

N96 Wath & Elsecar Disposal Point

N4 Barnburgh Colliery

N51-N55 Manvers Main

to N71

N80/N81 Skiers Spring North Disposal Point/ Colliery

N20/N21 Cortonwood Coking Plant & Colliery

N31 Elsecar Colliery

N94/N95 Wath Coking Plant & Colliery

to Sheffield

N30 Elsecar Workshops

N98 Wentworth Disposal Point

N81 Skiers Spring South Disposal Point

SWINTON

to Doncaster

N48 Kilnhurst Colliery

N97 Wentworth Drift Mine

N99 Wentworth Opencast Site

N93 Warren Vale Disposal Point

N101 Wentworth Stores

N3 Barley Hall Colliery

to Sheffield

to Rotherham

to Doncaster

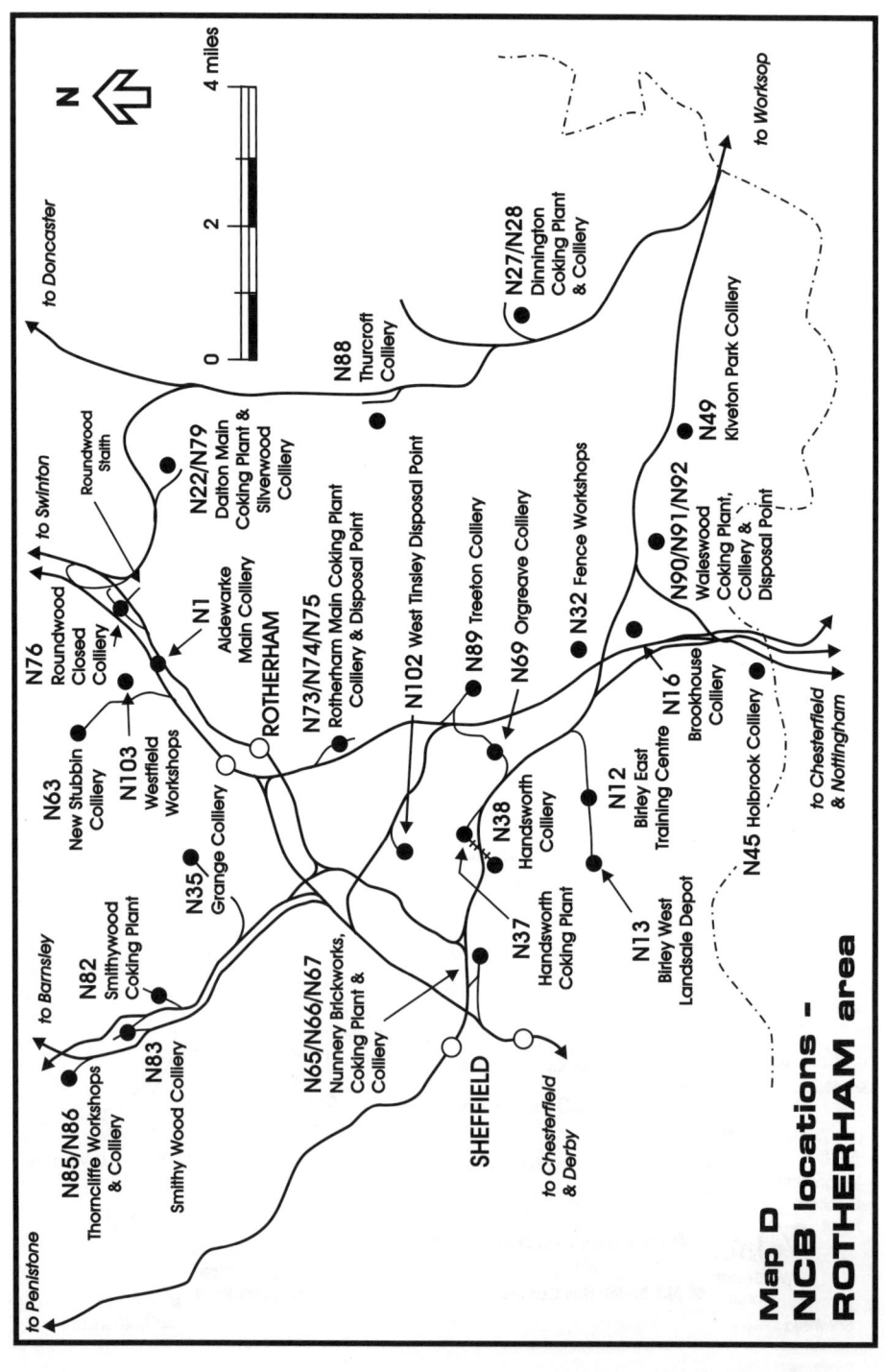

Map D
NCB locations –
ROTHERHAM area

N
4 miles
0 2

to Doncaster

to Swinton

Roundwood Staith

N22/N79 Dalton Main Coking Plant & Silverwood Colliery

N88 Thurcroft Colliery

N27/N28 Dinnington Coking Plant & Colliery

to Worksop

N76 Roundwood Closed Colliery

N1 Aldewarke Main Colliery

ROTHERHAM

N73/N74/N75 Rotherham Main Coking Plant & Disposal Point

N102 West Tinsley Disposal Point

N89 Treeton Colliery

N69 Orgreave Colliery

N32 Fence Workshops

N90/N91/N92 Waleswood Coking Plant, Colliery & Disposal Point

N49 Kiveton Park Colliery

N63 New Stubbin Colliery

N103 Westfield Workshops

Grange Colliery

N35 Smithywood Coking Plant

N38 Handsworth Colliery

N16 Brookhouse Colliery

N12 Birley East Training Centre

N45 Holbrook Colliery

to Chesterfield & Nottingham

to Barnsley

N82 Smithy Wood Colliery

N83

N85/N86 Thorncliffe Workshops & Colliery

Smithy Wood Colliery

N65/N66/N67 Nunnery Brickworks, Coking Plant & Colliery

N37 Handsworth Coking Plant

N13 Birley West Landsale Depot

SHEFFIELD

to Chesterfield & Derby

to Penistone

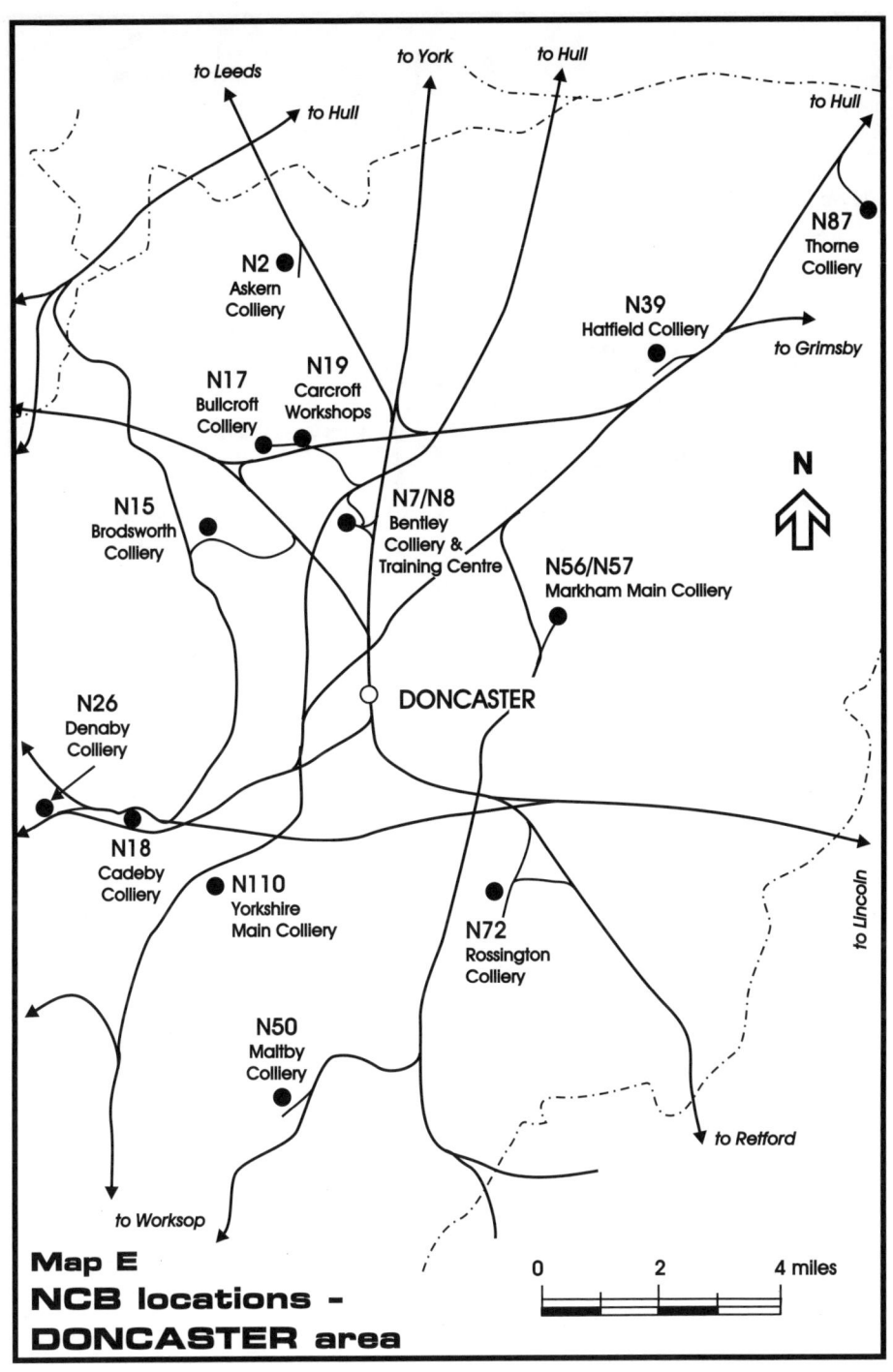

to Leeds

to York

to Hull

to Hull

to Hull

N87
Thorne
Colliery

N2
Askern
Colliery

N39
Hatfield Colliery

to Grimsby

N17
Bullcroft
Colliery

N19
Carcroft
Workshops

N

N15
Brodsworth
Colliery

N7/N8
Bentley
Colliery &
Training Centre

N56/N57
Markham Main Colliery

N26
Denaby
Colliery

DONCASTER

N18
Cadeby
Colliery

N110
Yorkshire
Main Colliery

N72
Rossington
Colliery

to Lincoln

N50
Maltby
Colliery

to Retford

to Worksop

0   2   4 miles

**Map E
NCB locations -
DONCASTER area**

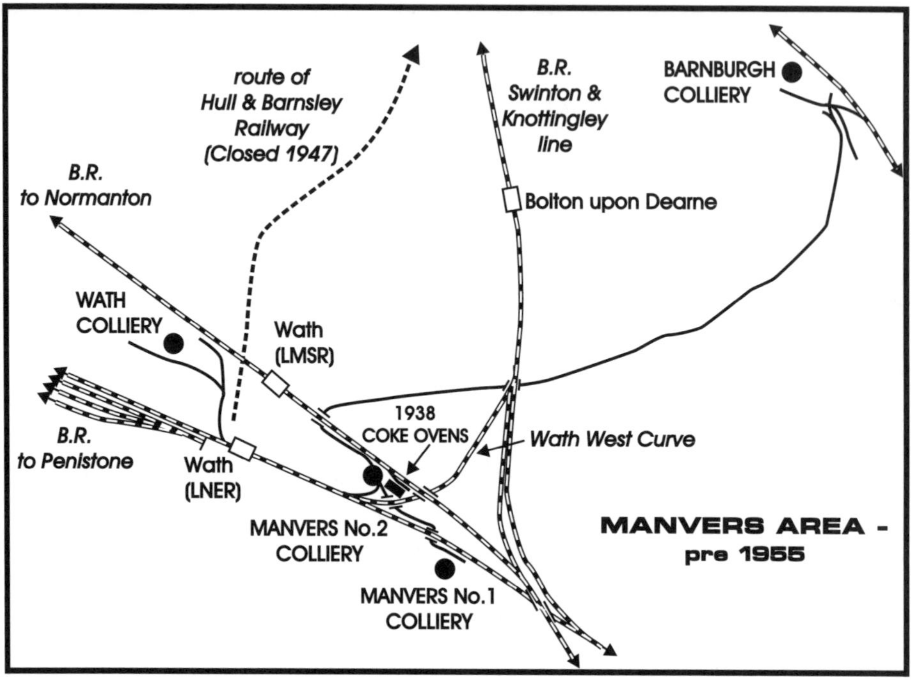

route of
Hull & Barnsley
Railway
(Closed 1947)

B.R.
Swinton &
Knottingley
line

BARNBURGH
COLLIERY

Bolton upon Dearne

B.R.
to Normanton

WATH
COLLIERY

Wath
(LMSR)

1938
COKE OVENS

Wath West Curve

B.R.
to Penistone

Wath
(LNER)

MANVERS No.2
COLLIERY

MANVERS No.1
COLLIERY

**MANVERS AREA -
pre 1955**

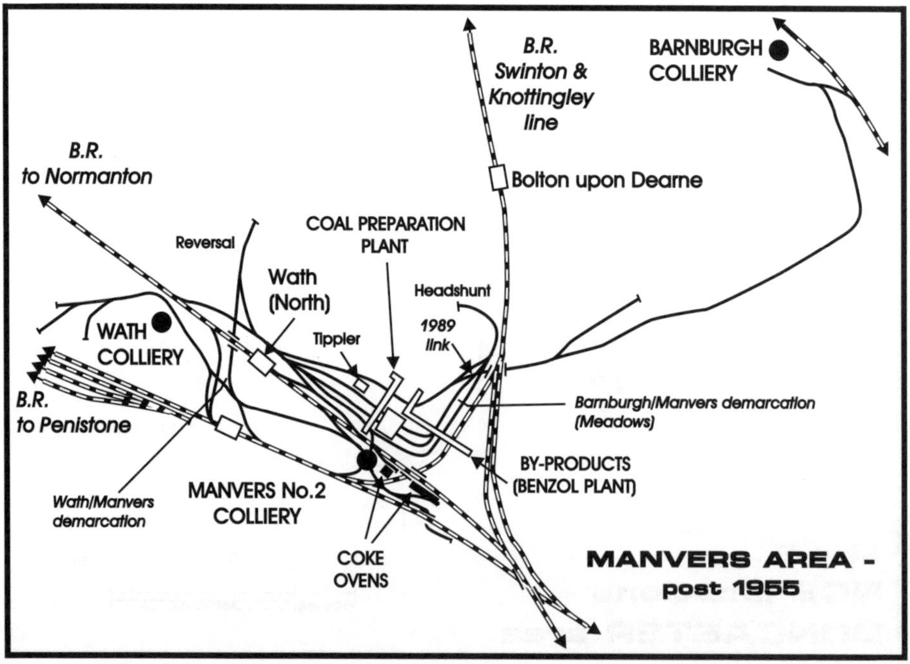

B.R.
Swinton &
Knottingley
line

BARNBURGH
COLLIERY

Bolton upon Dearne

B.R.
to Normanton

Reversal

COAL PREPARATION
PLANT

Headshunt

Wath
(North)

1989
link

WATH
COLLIERY

Tippler

Barnburgh/Manvers demarcation
(Meadows)

B.R.
to Penistone

BY-PRODUCTS
(BENZOL PLANT)

Wath/Manvers
demarcation

MANVERS No.2
COLLIERY

COKE
OVENS

**MANVERS AREA -
post 1955**

The colliery was served by sidings located south-east of the BR (ex LMSR) line, adjacent to Parkgate & Rawmarsh Station and north-west of the BR (ex LNER) line, south west of Parkgate & Aldwarke Station. A branch ran south-east, passing beneath the latter line, to Aldwarke Staith on the Sheffield & South Yorkshire Navigation (SK 442944). By 1958 the staith appears to have closed and the area was a rail served tip. Locomotives were used underground for coal haulage from 1949

Gauge : 4ft 8½in

No.4	ATLAS No.9	0-4-0ST	OC	HE	429	1887	(a)	(1)
	ATLAS No.1	0-4-0ST	OC	HE	475	1889	(a)	(2)
No.10	ATLAS No.15	0-4-0ST	OC	HL	2464	1900	(b)	(3)
	ATLAS No.7	0-4-0ST	OC	HE	311	1883	(c)	(4)
No.29	(RICHARD)	0-4-0ST	OC	MW	1968	1919	(d)	(5)
No.17		0-4-0ST	OC	HC	751	1906	(e)	(6)
No.8	(D.M.C.No.2)	0-4-0ST	OC	P	701	1898	(f)	(7)
No.4		0-4-0ST	OC	P	1114	1907	(g)	(8)
No.19		0-4-0ST	OC	HC	916	1910	(h)	(9)
	ATLAS No.6	0-4-0ST	OC	YE	478	1892	(j)	(10)
No.3		0-4-0ST	OC	YE	119	1869	(k)	(11)

(a) ex John Brown & Co Ltd, with site, 1/1/1947
(b) ex John Brown & Co Ltd, with site, 1/1/1947;
    to Elsecar Central Workshops, after 4/1958, by 2/1959; returned 26/2/1960
(c) ex Rotherham Main Colliery, c/1947
(d) ex Elsecar Central Workshops, 1/1949
(e) ex Roundwood Staith, after 8/1953, by 4/1954;
    to Elsecar Central Workshops, 7/1957; returned, after 7/1957, by 9/1958
(f) ex Silverwood Colliery, 10/6/1954;
    to Rotherham Main Colliery, after 6/1954, by 8/1954; returned 2/1955
(g) ex John Cashmore Ltd, Great Bridge, Staffordshire (West Midlands), hire, 12/1954
(h) ex Elsecar Central Workshops, 6/2/1958
(j) ex New Stubbin Colliery, 3/11/1958
(k) ex Elsecar Colliery, after 3/1959, by 9/1959

(1) scrapped by Thos. W. Ward Ltd, 4/1957
(2) scrapped, /1950 (after 27/5/1950)
(3) to Silverwood Colliery, c11/1962 (after 3/1962, by 7/1963)
(4) to Rotherham Main Colliery, c/1949 (by 15/4/1949)
(5) scrapped on site by Frank Tingle Ltd of Kilnhurst, 9/1961
(6) to Denaby Colliery, 9/1961
(7) sold or scrapped, c11/1962 (after 10/1962)
(8) to Manvers Colliery (still on hire), 5/7/1955
(9) to Elsecar Colliery, 6/8/1958
(10) to New Stubbin Colliery, 7/1960
(11) to New Stubbin Colliery, after 3/1959, by 9/1959

Gauge : 1ft 11in (Underground locomotives)

No.5		4wDMF	RH	268860	1949	New	(1)
-		4wDMF	RH	268868	1950	New	(2)
No.6		4wDMF	RH	249567	1947	(a)	(2)
No.5		4wDMF	RH	256275	1948	(a)	(3)

(a) ex Elsecar Central Workshops, after 5/1956, by 1/6/1957 (converted from 2ft 0in gauge)

(1)   to Elsecar Central Workshops, 5/1961
(2)   to Barnburgh Colliery, 7/1962
(3)   scrapped on site by Frank Tingle Ltd of Kilnhurst, 7/1961

ASKERN COLLIERY, Askern   N2
ex Askern Coal & Iron Co Ltd   SE 557139
NE2 from 1/1/1947; DCR from 26/3/1967; NYK from 1/10/1985; NYG from 1/4/1990; SEL from
1/10/1991.   CLOSED 12/1991

Sidings ran south-west from the BR (ex LNER) line, ¼ mile south of the closed Norton Station, to the
colliery (¾ mile). Branches ran north-west to extensive dirt tips (½ mile) and there was a connection
to the adjacent works of Doncaster Coalite Ltd. The use of standard gauge locomotives ceased in
1980, after rapid loading of BR trains commenced. Use of underground locomotives commenced in
1956 but their purpose is not known.

Reference : "Askern Coalite Works" – articles in "Industrial Railway Records" Nos.103 & 106

Gauge : 4ft 8½in

	MORECAMBE	0-6-0ST	IC	MW	1393	1898	(a)	Scr 4/1952
No.22	VICTORY	0-6-0ST	OC	AE	1834	1919	(a)	(1)
	LITTLETON	0-6-0ST	OC	AE	1833	1919	(a)	(2)
No.6	CHESTERFIELD	0-6-0ST	IC	MW	1667	1906	(b)	(3)
No.45	AM218 3219/002	0-6-0DE		RH	384146	1956	New	(4)
No.50	68020	0-6-0ST	IC	WB	2752	1944	(c)	(5)
RM893	ROSSINGTON No.1	0-6-0ST	IC	HE	3594	1950	(d)	(6)
No.35	YM/12/51/M (75050)	0-6-0ST	IC	RSHN	7086	1943	(e)	(7)
D2598	SAM	0-6-0DM		HE	5647	1960	(f)	(8)
RM2020	TOMMY 3219/020	0-6-0DH		HC	D1386	1966	(g)	(9)
D2599	F.SE 357 3219/013	0-6-0DM		HE	5648	1960	(h)	(10)

(a)   ex Askern Coal & Iron Co Ltd, with site, 1/1/1947
(b)   ex Yorkshire Main Colliery, after 5/1951, by 8/1952
(c)   ex BR, 15/6/1963
(d)   ex Rossington Colliery, 8/6/1968
(e)   ex Yorkshire Main Colliery, c27/4/1970
(f)   ex Rossington Colliery, 6/1971 or 7/1971
(g)   ex Markham Main Colliery, 2-5/10/1974
(h)   ex Frickley Colliery, West Yorkshire, 16/6/1976

(1)   to Bentley Colliery, c/1961 (after 9/1961)
(2)   scrapped, after 10/1968, by 3/1969
(3)   scrapped, after 31/3/1956, by 4/1957
(4)   scrapped by R.D. Geeson, of Ripley, Derbyshire, 6/1981
(5)   scrapped on site by George Cohen, Sons & Co Ltd, 5/1970
(6)   boiler to Steamport, Southport, Merseyside, remains scrapped on site by Mee & Cocker Ltd
       of Leigh, Lancashire, 3/1977
(7)   written off, 7/6/1976; to Titanic Salvage Co Ltd, Ellastone, Staffordshire,. 24-26/1/1977;
       later Kent & East Sussex Railway, Tenterden, Kent
(8)   fire damaged 1974; to Philadelphia Central Workshops, Co. Durham 2/1975 and scrapped
       there 5/1975 (after 16/5/1975)
(9)   to Hatfield Colliery, after 28/7/1980, by 1/1981
(10)   scrapped by R.D. Geeson, of Ripley, Derbyshire, 5/1981

Note that 200HP 0-6-0DM locomotives were regularly hired from British Railways and
frequently changed during the years 1969 - 1971 and 1973 – 1976.

Gauge 2ft 0in (Surface stockyard)

-	4wDM	RH	223699	1944	(a)	(1)	
390/A/M/2491	4wDMF	RH	268857	1948	(b)	(2)	
-	0-4-0DMF	RH	370543	1954	(b)	s/s c/1971	
390/MM/M/2521	0-4-0DMF	HC	DM749	1949	(c)	(3)	

(a)   ex unknown location, by 7/1960 (originally new to MoS)
(b)   ex Thorne Colliery c/1958
(c)   ex underground, 11/2/1970

(1)   sold or scrapped c/1966 (after 10/1964, by 10/1968)
(2)   written off, 26/6/1974, and scrapped soon afterwards
(3)   to Leeds Industrial Museum, Armley Mills, West Yorkshire, after 10/3/1992, by 18/6/1992

Gauge : 2ft 0in (Underground locomotives)

	390/A/M/2083	0-6-0DMF	HC	DM931	1956	New	(1)
	390/A/M/2125	0-6-0DMF	HC	DM987	1956	New	(1)
	390/A/M/4119	0-6-0DMF	HC	DM1107	1959	New	(1)
	390/MM/M/2521	0-4-0DMF	HC	DM749	1949	(a)	(2)
No.15	390/HA/M/07 18 No.112	0-6-0DMF	HC	DM718	1951	(b)	(1)
No.11	390/RM/2120	0-6-0DMF	HC	DM802	1954	(c)	(1)
	390/R/M/2228	0-6-0DMF	HC	DM937	1956	(d)	(1)
No.15	390/R/M/2168	0-6-0DMF	HC	DM930	1955	(e)	(1)
No.8	390/R/M/2084	0-6-0DMF	HC	DM800	1953	(f)	(1)
	-	4wBEF	CE	B3269	1986	(g)	(3)
No.10	F/SE 1228	0-6-0DMF	HE	7482	1978	(h)	(4)
No.11	F/SE 1229	0-6-0DMF	HE	7483	1978	(j)	(5)
No.12	F/SE 1230	0-6-0DMF	HE	8578	1978	(j)	(5)
No.13	F/SE 1231	0-6-0DMF	HE	8579	1978	(j)	(5)
No.28		0-6-0DMF	HC	DM1331	1964	(k)	(6)
No.25		0-6-0DMF	HC	DM1380	1966	(k)	(5)
No.26		0-6-0DMF	HC	DM1393	1967	(k)	(6)
No.27		0-6-0DMF	HC	DM1395	1966	(k)	(5)

(a)   ex Carcroft Central Workshops, 7/12/1962
(b)   ex Hatfield Colliery, 11/1968
(c)   ex Rossington Colliery, 2/2/1970
(d)   ex Rossington Colliery, 12/1/1971
(e)   ex Rossington Colliery, 14/1/1971
(f)   ex Rossington Colliery, 20/10/1971
(g)   ex Barnburgh Colliery (2ft 2in gauge), 28/7/1989;
        to CE, Hatton, Derbyshire, 8/8/1989, and altered to 2ft 0in gauge;
        returned /1990 (by 9/3/1990)
(h)   ex Frickley Colliery, West Yorkshire, 8/10/1987
(j)   ex Frickley Colliery, West Yorkshire, 8/10/1987; re-gauged from 2ft 1in gauge
(k)   ex Calverton Colliery, Nottinghamshire, 8/1991 (not used underground)

(1)   sold, scrapped or abandoned underground by /1992
(2)   to surface stockyard, 11/2/1970
(3)   to Sharlston Colliery, West Yorkshire, /1992
(4)   re-gauging from 2ft 1in never completed; loco still on surface 2/4/1990;
        sold or scrapped by /1992
(5)   sold or scrapped by /1992
(6)   to Rossington Colliery, 26/6/1992

BARLEY HALL COLLIERY, Thorpe Hesley

SK 373966

Drift opened by NE5 c1947 and production ceased c1949; colliery opened 1955;
BNY from 26/3/1967.

CLOSED 5/1974

This site was an air shaft for THORNCLIFFE COLLIERY and about 1947 a short-term drift was driven nearby. In 1955 the air shaft and surface installations were redeveloped as a colliery whose coal was wound at SMITHYWOOD COLLIERY. No standard gauge rail connection and locomotives were not used underground.

BARNBURGH COLLIERY, Barnburgh

N4

ex **Manvers Main Collieries Ltd**

SE 475033

NE3 from 1/1/1947; SYK from 26/3/1967.

CLOSED 16/6/1989

The colliery was served by sidings on the west side of the BR (ex LMSR) Dearne Valley Railway, ½ mile north-west of Harlington Halt. A NCB line ran south-west from the colliery to MANVERS COLLIERY (2½ miles). After the opening of the MANVERS CENTRAL COAL PREPARATION PLANT in 1956, all traffic went by this route. The Barnburgh screens closed and the siding agreement with BR at Barnburgh was terminated from 21/7/1958. Locomotives were used underground on two gauges, the 3ft 0in gauge to bring coal to the shaft and the 2ft 2in for manriding and materials.

Reference : A.J. Booth, *"Manvers Main and Barnburgh Main"* (IRS, 1996)

Gauge : 4ft 8½in

In addition to the locomotives listed below, 350HP diesel shunters were hired from British Railways from 1986 until closure. These hired locos were exchanged frequently.

No.5		0-6-0ST	IC	P	1242	1911	(a)	(1)
42	MANVERS MAIN No.11	0-6-0T	IC	HC	1690	1937	(b)	(2)
	MANVERS MAIN No.6	0-6-0ST	OC	HC	822	1912	(c)	(3)
48		0-6-0ST	IC	HE	3685	1950	New (d)	(4)
49	(No.15)	0-6-0ST	IC	HE	3701	1950	New (e)	(5)
(No.28)		0-6-0ST	OC	HC	1364	1919	(f)	(6)
No.50		4wVBT	VCG	S	9552	1952	(g)	(7)
No.41	ELSIE	0-6-0ST	OC	WB	2223	1924		
				reb		1937	(h)	(8)
(HAROLD No.57)	KERRY No.57	0-6-0DM		RH	347748	1958	New (j)	(9)
	GEOFFREY No.60	0-6-0DM		RH	347749	1958	New (k)	(10)
39	FREDERICK	0-6-0T	OC	HL	3676	1927	(m)	(11)
No.52	DENNIS	0-6-0ST	IC	HE	3832	1955	(n)	(12)
No.2	HARRY No.73	0-6-0DH		HE	6661	1966	New (p)	(13)
(D2373)	DAWN No.1 (JIM)	0-6-0DM		Sdn		1962	(q)	(14)
(D2337)	DOROTHY No.3	0-6-0DM		RSHD	8196	1961		
				DC	2718	1961	(r)	(15)
D2225	DEBRA	0-6-0DM		VF	D274	1955		
				DC	2548	1955	(s)	(16)

(a)   ex Manvers Main Collieries Ltd, with site, 1/1/1947
(b)   ex Manvers Main Collieries Ltd, with site, 1/1/1947;
         to Manvers Colliery, /1953 (after 28/8/1950); returned, after 4/1953, by 8/1953
(c)   ex Manvers Colliery, after 9/1947, by 10/1949
(d)   to Manvers Colliery, after 7/1958, by 3/1959; returned 8/1960 or 9/1960
(e)   to Manvers Colliery, after 4/1951, by 6/1952; returned, after 8/1953, by 29/5/1955
(f)   ex Wath Colliery, after 1/1953, by 8/1953
(g)   ex Manvers Colliery, /1954

(h) ex Manvers Colliery, after 4/1957, by 4/1958;
     to Manvers Colliery, after 7/1958, by 7/1959; returned after 11/1959, by 9/1960
(j) to Manvers Colliery, after 16/5/1970, by 8/1970; returned, 17/6/1974;
     to Manvers Colliery, 24/6/1974; returned, 5/10/1978;
     to Manvers Colliery, 18/10/1978; returned, 25/11/1983
(k) to Manvers Colliery, after 5/1969, by 7/1971; returned, 4/4/1973;
     to Manvers Colliery, 9/7/1973; returned, by 9/1973;
     to Manvers Colliery, after 6/1974, by 11/1975; returned, 20/12/1978;
     to Manvers Colliery, 11/1/1979; returned, 15/1/1979;
     to Manvers Colliery, 18/1/1979; returned, 1/7/1983
(m) ex Silverwood Colliery 31/5/19590;
     to Manvers Colliery, after 8/1959, by 10/1959; returned after 3/1964, by 5/1964
(n) ex Manvers Colliery, 13/5/1961;
     to Manvers Colliery, 7/1961 (by 15/7/1961); returned, after 7/1963, by 3/1964;
     to Manvers Colliery, after 3/1964, by 11/1964; returned, after 3/1966, by 4/1966
(p) to HE, Leeds, West Yorkshire, for repairs, 9/1/1974, and returned, 26/7/1974
(q) ex Manvers Colliery, 7/1971;
     to Manvers Colliery, 22/10/1975; returned, after 10/1975, by 3/1976
(r) ex Manvers Colliery, 6/1974
(s) ex Manvers Colliery, 16/3/1976

(1) to Manvers Colliery, by 18/4/1949
(2) to Manvers Colliery, after 8/1953, by 7/1955
(3) to Manvers Colliery, after 10/1949, by 5/1950
(4) to Manvers Colliery, after 9/1960, by 12/3/1961
(5) to Manvers Colliery, after 4/1958, by 7/1958
(6) scrapped on site by F. Green, of Stairfoot, 8/1967
(7) to Wath Colliery, 8/1955
(8) to Manvers Colliery, after 9/1960 by 3/1961
(9) scrapped, after 17/7/1986, by 12/12/1986
(10) to C.F. Booth Ltd, Rotherham, 13/3/1986, and scrapped there, 21/3/1986
(11) to Manvers Colliery after 5/1964, by 11/1964
(12) to Manvers Colliery after 4/1966, by 1/1967
(13) to Manvers Colliery after 15/5/1986, by 7/7/1986
(14) to Manvers Colliery, 3/1976 or 4/1976
(15) to Manvers Colliery, 2/1977
(16) to Manvers Colliery, after 3/1976, by 6/1976

Gauge : 3ft 0in (Underground Locomotives)

No.17	390/3314	0-6-0DMF	HE	4036	1949	New	(1)
No.19	390/3315	0-6-0DMF	HE	4072	1954	New	(1)
No.20	390/3316	0-6-0DMF	HE	4073	1954	New	(1)
	-	0-6-0DMF	HE	3431	1947	(a)	(2)

(a) ex Manvers Colliery (date unknown)

(1) abandoned underground 6/1989
(2) written off by 7/1978

Gauge : 2ft 2in (Underground locomotives)

	-	0-4-0DMF	HE	3316	1946	(a)	(1)
	-	4wDMF	RH	268868	1950	(b)	(1)
No.6		4wDMF	RH	249567	1947	(b)	(1)
No.61	390/3541	0-6-0DMF	HE	6227	1963	New	(2)
No.62	390/3558	0-6-0DMF	HE	6228	1963	New	(2)

No.72	390/2685	0-4-0DMF		HE	4130	1950		
			reb	HE	7222	1970	(c)	(2)
No.73	390/7223	0-4-0DMF		HE	#	1957		
			reb	HE	7223	1971	(d)	(2)
--		4wBEF		CE	B3269	1986	New	(3)

works number either 4332 or 5203 – see note under Cadeby Colliery

(a) ex Manvers Colliery (date unknown)
(b) ex Aldwarke Main Colliery (converted from 1ft 11in gauge), 7/1962
(c) ex HE, Leeds, West Yorkshire, 5/11/1971 (earlier 1ft 11in gauge at Cadeby Colliery)
(d) ex Fence Central Workshops, 22/7/1971

(1) written off by 7/1978
(2) abandoned underground 6/1989
(3) to Askern Colliery, 28/7/1989

Gauge : 2ft 0in (spare parts donor for underground locomotives)

No.55	0-4-0DMF	HE	6059	1962	(a)		(1)

(a) ex Elsecar Colliery, 3/1980

(1) remains scrapped, /1980

BARNSLEY MAIN COLLIERY, Barnsley   N5
ex **Barrow Barnsley Main Collieries Ltd**   SE 365061

NE5 from 1/1/1947 Colliery & rail system CLOSED 5/1966; Colliery re-opened by BNY c1984; NYK from 1/10/1985.   CLOSED 7/1991

The colliery (which included the surface installations of the adjacent closed OAKS COLLIERY, where the locoshed was sited) was located in the complex of BR lines north of the ex LNER line between Stairfoot and Barnsley Court House stations. The rail connection was to this line. A former connection to the ex LMS Barnsley to Cudworth line may have closed prior to vesting day. The rail connection was removed after the 1966 closure and not reinstated. The underground locomotive actually was used on the surface and was flame-proofed to permit it to enter the pit top air-lock.

Gauge :: 4ft 8½in

5	(2)	0-4-0ST	OC	AE	1838	1919		
		reb		YE		1927	(a)	Scr 5/1954
(No.1)		0-6-0ST	IC	P	1518	1919	(b)	(1)
	-	0-6-0ST	OC	YE	1889	1923	(c)	(2)
	-	4wVBT	VCG	S	9398	1950	(d)	(3)
	SENTINEL No.1	4wVBT	VCG	S	9400	1950	New	(4)
	SENTINEL No.3	4wVBT	VCG	S	9394	1950	(e)	(5)
	H.C.No.4	0-4-0ST	OC	HC	1892	1961	New	(6)
	YORK No.1	0-4-0ST	OC	YE	2474	1949	(f)	(7)
	-	0-4-0DM		JF	22558	1939	(g)	(8)

(a) ex Barrow Barnsley Main Collieries Ltd, with site, 1/1/1947;
    (may have been at Barrow Colliery for a period between 8/1950 and 5/1951)
(b) ex Barrow Barnsley Main Collieries Ltd, with site, 1/1/1947;
    to Barrow Colliery after 4/1949, by 18/9/1949; returned, /1957 (after 4/1957)
(c) ex Barrow Barnsley Main Collieries Ltd, with site, 1/1/1947;
    to Thorncliffe Central Workshops, after 8/1950, by 7/1951;
    returned, after 29/6/1952, by 1/7/1953
    to Birdwell Central Workshops after 4/1957 by 7/1958; returned, after 7/1958, by 6/1959

(d)   earlier on demonstration at Edgar Allen & Co Ltd, Sheffield, by 21/3/1950;
to here on demonstration, 3/1950 (by 21/3/1950)
(e)   ex Barrow Colliery, after 7/1951, by 29/6/1952;
to Skiers Spring Colliery after 6/1952, by 28/6/1953; returned 6/8/1953
(f)   ex Monk Bretton Colliery, 2/1966
(g)   ex Wharncliffe Silkstone Colliery, 4/1966 or 5/1966

(1)   to Barrow Colliery, after 10/1959, by 3/1961
(2)   scrapped, after 1/1962, by 5/1962
(3)   later at Kirkby Colliery, Nottinghamshire, by 4/1950
(4)   to Wombwell Colliery, 20/3/1964
(5)   scrapped on site, /1968 (after 4/1968)
(6)   to Wombwell Colliery after 5/1966, by 4/1967
(7)   to Monk Bretton Colliery, after 2/1966, by 5/1966
(8)   to Dearne Valley Colliery, after 7/1967, by 4/1968

Gauge : 2ft 0in (Surface locomotive)

--	4wBEF	CE	B3086	1984	New	(1)

(1)   to Thyssen Tunnelling Ltd, contractors, Pontefract, West Yorkshire, after 7/1991, by 3/9/1992

BARROW COLLIERY, Worsbrough   N6
ex Barrow Barnsley Main Collieries Ltd   SE 359025
NE5 from 1/1/19; BNY from 26/3/1967.   Rail traffic ceased c/1985;
Merged with BARNSLEY MAIN c/1986 and surface CLOSED

The colliery was located at the end of a ¾ mile branch which ran west from the BR (ex LNER) line, south of Dovecliffe Station and was connected at the same point to the parallel ex LMS line. The branch also served the Barrow works of Barnsley District Coking Co Ltd, located on the south side, midway between colliery and main line. The duties of the underground locomotives are not known.

Gauge : 4ft 8½in

(No.4)		0-4-0ST	OC	P	1627	1924	(a)	(1)
No.3		0-4-0ST	OC	WB	2105	1919	(a)	(2)
	RAYMOND	0-4-0ST	OC	HC	810	1907	(b)	(3)
(75131)	3181	0-6-0ST	IC	HE	3181	1944	(c)	(4)
	(YORK No.2)	0-4-0ST	OC	YE	2473	1949	New (d)	(5)
	VENTURE	0-4-0DM		JF	22287	1938	(e)	(6)
(No.1)		0-6-0ST	IC	P	1518	1919	(f)	(7)
	SENTINEL No.3	4wVBT	VCG	S	9394	1950	(g)	(8)
	-	0-4-0ST	OC	AE	1839	1919		
		reb		YE		1927	(h)	(9)
	SENTINEL No.5	4wVBT	VCG	S	9570	1954	New	(10)
	SENTINEL No.6	4wVBT	VCG	S	9616	1957	(j)	(11)
	AVON No.3	0-6-0ST	OC	AE	1826	1919	(k)	(12)
	H.C.No.5	0-4-0ST	OC	HC	1893	1961	New	(13)
(TL24)	TL40	4wDH		TH	156C	1965		
	built incorporating the chassis of S				9570		(m)	(14)
2219		0-6-0DM		VF	D268	1955		
				DC	2542	1955	(n)	(15)
(TL3)	TL39	4wDH		TH	158C	1965		
	built incorporating the chassis of S				9617		(p)	(16)
DH17		4wDH		S	10176	1964	(q)	(14)
	-	0-4-0DE		YE	2729	1958	(r)	(17)

D2199 ROCKINGHAM COLLIERY No.1

                     0-6-0DM   Sdn   1961   (s) (18)

(a)   ex Barrow Barnsley Main Collieries Ltd, with site, 1/1/1947
(b)   ex Thorncliffe Central Workshops, /1948
(c)   ex WD, after return from Belgium, by 18/9/1949;
       to Birdwell Central Workshops, after 1/1962, by 4/1962; returned 19/3/1964
(d)   to Monk Bretton Colliery after 9/1953, by 16/5/1954; returned 19/1/1955
       to Birdwell Central Workshops c/1960 (by 12/3/1961); returned, after 14/10/1961, by 3/1964
(e)   ex Grange Colliery, after 1954, by 4/1955
(f)   ex Barnsley Main Colliery, after 4/1949, by 18/9/1949;
       to Barnsley Main Colliery, /1957 (after 4/1957); returned after 10/1959, by 3/1961
(g)   ex Monk Bretton Colliery after 24/12/1950, by 19/5/1951
(h)   ex Barnsley District Coking Co Ltd, (hire?) after 4/1957, by 7/1958
(j)   ex Smithy Wood Colliery, after 12/1957, by 16/2/1958
(k)   ex Dodworth Colliery, after 11/1959, by 12/3/1961
(m)   ex Smithy Wood Colliery, /1965 (by 23/1/1966)
(n)   ex Barnsley District Coking Co Ltd, Barrow Coking Plant, Worsbrough, hire, c8/1969
(p)   ex Dodworth Colliery, after 8/1970, by 3/1971
(q)   ex TH, by rail, 25/3/1971
(r)   ex Barnsley District Coking Co Ltd, Barrow Coking Plant, Worsbrough, hire, c10/1976
(s)   ex Rockingham Colliery after 10/1978, by 4/1979;
       to Houghton Main Colliery, after 5/1979, by 9/1979;
       ex Royston Drift Mine, West Yorkshire, 8/7/1981

(1)   to Skiers Spring Colliery after 7/1958, by 10/1959
(2)   scrapped by Thos. W. Ward Ltd, /1952
(3)   to Wharncliffe Silkstone Colliery, after 9/1949, by 1/1953
(4)   scrapped on site by unidentified merchant, of Shafton, c1/2/1969
(5)   to Grimethorpe Colliery, /1966 (after 4/1966)
(6)   to Birdwell Central Workshops, after 5/1955, by 9/1955
(7)   to Dodworth Colliery, after 11/1964, by 4/1965
(8)   to Barnsley Main Colliery, after 7/1951, by 6/1952
(9)   returned to Barnsley District Coking Co Ltd, after 7/1958, by 10/1959
(10)   to Smithy Wood Colliery after 4/1957, by 16/2/1958
(11)   to Smithy Wood Colliery, 3/1960
(12)   to Wombwell Colliery, 22/3/1964
(13)   scrapped on site, after 6/1972, by 2/1973
(14)   to C.F. Booth Ltd, Rotherham, 2/4/1986, and scrapped there, c6/10/1987
(15)   returned to Barnsley District Coking Co Ltd, Barrow Coking Plant, c8/1969
(16)   to South Kirkby Colliery, West Yorkshire, after 7/1971, by 9/1971
(17)   returned to Barnsley District Coking Co Ltd, Barrow Coking Plant, c/1976
(18)   to Royston Drift Mine, West Yorkshire, 23/3/1982

Gauge : 2ft 0in (Underground locomotives)

		4wBEF	Bg	3400	1953		
-			MV	880	1953	New	Scr c/1971
-		4wBEF	Bg	3401	1953		
			MV	881	1953	New	Scr c/1971

NE2 from 1/1/1947; DCR from 26/3/1967: SYK from 1/10/1985.   CLOSED 12/1993

The colliery was located at the end of a ½ mile line which ran west from the BR (ex LNER) Doncaster to York line, ½ mile north of the closed Arksey Station. It was, until after 1960, also connected to the former GC and Hull & Barnsley joint line north west of the colliery at Bullcroft Junction. A line also ran south for ½ mile to SE 572063, where there is believed to have been a landsale yard. Use of standard gauge locomotives ceased from c1976, after the introduction of rapid loading. Locomotives were used extensively underground for all purposes.

Gauge : 4ft 8½in

(BENTLEY) BENTLEY No.3		0-6-0T	IC	HC		850	1909		
		rebuilt		Carcroft CW	#	1963	(a)	(1)
	BENTLEY No.2	0-6-0T	IC	HC		851	1909	(b)	(2)
	BENTLEY No.3	0-6-0T	IC	HC		925	1910	(b)	(2)
75023		0-6-0ST	IC	HE		2872	1943	(c)	(3)
(70)	10	0-6-0ST	OC	HC		1348	1918	(d)	(4)
No.1	3219/004	0-6-0DM		HC	D1086	1958	New	(5)
No.22	VICTORY	0-6-0ST	OC	AE		1834	1919	(e)	(6)
D2613	BRM 5481 3219/008	0-6-0DM		HE		5662	1960	(f)	(7)

#	rebuild, which started in 1962, made use of component salvaged from HC 851 and HC 925.
(a)	ex Barber Walker & Co Ltd, with site,1/1/1947;
	to Carcroft Central Workshops, after 7/1960, by 5/1961; returned c6/1963 (by 7/1963)
(b)	ex Barber Walker & Co Ltd, with site,1/1/1947
(c)	ex War Department, after return from hire to Belgian State Railways, /1948;
	to Carcroft Central Workshops, /1963 (by 7/1963); returned, after 6/1964, by 3/1965
(d)	ex Appleby Frodingham Steel Co Ltd, Scunthorpe, Lincolnshire, 7/1950
(e)	ex Askern Colliery, c/1961 (after 9/1961)
(f)	ex Brodsworth Colliery, after 12/12/1973, by 8/1975

(1)	scrapped on site by Thos.W.Ward Ltd, 12/1973
(2)	dismantled /1957, some parts kept for use on repair of HC 850, remains scrapped 8/1957
(3)	written off, 20/2/1975; scrapped on site by NCB, after 8/1975, by 10/1975
(4)	sold or scrapped, after 7/1963, by 12/1963
(5)	written off, 6/4/1977; scrapped on site by Walter Heslewood Ltd, 9/1977
(6)	scrapped c6/1962 (by 7/1963)
(7)	written off, 6/4/1977; scrapped on site by Walter Heslewood Ltd, 6/1977

Gauge : 3ft 0in (here as source of spare parts only)

390/YM/1675/M	0-4-0DMF	HE	3615	1948	(a)	(1)
390/YM/1691/M	0-4-0DMF	HE	3617	1948	(a)	(1)

(a)	ex Yorkshire Main Colliery, c/1986 (by 18/5/1986)

(1)	remains scrapped /1991

Gauge : 2ft 6in (here as source of spare parts only)

-	0-6-0DMF	HC	DM1140	1959	(a)	(1)

(a)	ex Allerton Bywater Colliery, West Yorkshire, 30/6/1992

(1)	to Embsay Steam Railway, Embsay, North Yorkshire, /1994

Gauge : 2ft 3½in (Surface stockyard)

BENTLEY No.2	390/BE/M/450	0-4-0DMF	HE	2662	1941	(a)	(1)
No.10	390/BE/M/19638	0-4-0DMF	HE	3573	1947	(b)	Scr c8/1993

(a)   ex underground, after /1965, by 10/1968
(b)   ex underground, 5/1968

(1)   scrapped (by Thos.W.Ward Ltd ?), /1973

Gauge : 2ft 3½in (Underground locomotives)

-		0-4-0DMF	HE	2002	1939	(a)	(1)
-		0-4-0DMF	HE	2661	1941	(a)	(2)
BENTLEY No.2	390/BE/M/450	0-4-0DMF	HE	2662	1941	(b)	(3)
No.3	390/BE/M/482	0-4-0DMF	HE	2663	1941	(c)	(4)
No.7	No.5 390/BE/M/514	0-4-0DMF	HE	3437	1946	(a)	(5)
No.8	No.6 390/BE/M/545	0-4-0DMF	HE	3438	1946	(a)	(5)
No.9	No.7 390/BE/M/576	0-4-0DMF	HE	3439	1946	(a)	(5)
No.11	No.8 390/BE/M/607	0-4-0DMF	HE	3440	1946	(a)	(5)
-		4wDMF	RH	249567	1947	New	(6)
No.10	390/BE/M/638	0-4-0DMF	HE	3573	1947	New	(7)
No.11	390/BE/M/669	0-4-0DMF	HE	3574	1947	New	(8)
No.12	390/BE/M/700	0-4-0DMF	HE	3575	1947	New	(8)
No.13	390/BE/M/731	0-4-0DMF	HE	3576	1947	New	(8)
No.14	390/BE/M/762	0-4-0DMF	HE	3577	1947	New	(8)
No.15	390/BE/M/793	0-4-0DMF	HE	3578	1947	New	(8)
No.16	390/BE/M/824	0-4-0DMF	HE	4494	1953	New	(8)
No.17	390/BE/M/1920	0-4-0DMF	HE	5205	1957	New	(8)
No.18	390/BE/M/1950	0-4-0DMF	HE	5206	1957	New	(8)
No.19	390/BE/M/1976	0-6-0DMF	HE	7480	1977	New	(8)
No.1	BE/M/1952	4wBEF	CE	B3271A	1985	New	(9)
No.2	BE/M/1953	4wBEF	CE	B3271B	1985	New	(9)
	524/50	4wBEF	CE	B2966B	1982		
	repaired	4wBEF	CE	B3568	1989	(d)	(8)
-		4w-4wBEF	CE	B3656	1990	New	(10)
-		4wBEF	CE	B3038	1983	(e)	(8)

(a)   ex Barber Walker & Co Ltd, with site,1/1/1947
(b)   ex Barber Walker & Co Ltd, with site,1/1/1947;
        to Carcroft Central Workshops, 6/2/1964; returned c/1965
(c)   ex Barber Walker & Co Ltd, with site,1/1/1947;
        to Carcroft Central Workshops, 9/9/1961; returned 16/10/1964
(d)   ex Treeton Colliery, by /1989;
        to CE, Hatton, Derbyshire, (altered from 1ft 10 in gauge); returned, 2/1989
(e)   ex CE, Hatton, Derbyshire, (altered from 2ft 0 in gauge), /1992;
        earlier at Dinnington [South Yorkshire] Colliery

(1)   written off, by c/1971
(2)   scrapped, 6/1965
(3)   to surface stockyard, after /1965, by 10/1968
(4)   scrapped (by Thos.W.Ward Ltd ?), /1973
(5)   written off by 10/1988
(6)   to Wentworth Drift Mine, 11/1949
(7)   to surface stockyard, 5/1968
(8)   sold, scrapped or abandoned underground, by /1994
(9)   transferred, sold, scrapped or abandoned underground, by /1994
(10) to Kellingley Colliery, North Yorkshire, /1994

Gauge : 2ft 0in (used on contract ?)

		4wDM	FH	1941	1934	(a)	s/s)

(a) ex unknown location, c/1954 (by 2/5/1955)
     (originally Thames Conservancy, London.)

BENTLEY TRAINING CENTRE, Bentley

Opened by DCR c1976; SYK from 1/10/1985; SYG from 1/4/1990.   CLOSED 1993

Located within the Bentley colliery complex, this comprised a steeply graded narrow gauge track used for training underground locomotive drivers.

Gauge : 3ft 0in

2	390/BR/M/774	(No.1)	0-6-0DMF	HC	DM1120	1958	(a)	(1)
No.1	390/BR/M/775	BEM401	0-6-0DMF	HC	DM914	1958	(b)	Scr /1989
No.2	390/YM/1668/M	BEM402	0-4-0DMF	HE	3614	1948	(c)	(1)
No.3	390/14501	BEM403	0-6-0DMF	HE	4816	1955	(d)	(1)
		BEM404	4wDHF rack	HE	8505	1981	(e)	(1)
	-		4w-4wBEF	CE	B3352A	1987		
		repaired		CE	B3478	1988	(f)	(2)

(a) ex Hickleton Training Centre, 5/1976;
     to Ashington Central Workshops, Northumberland, c/1980; returned by 18/2/1980
(b) ex Brodsworth Colliery, 9/1/1979
(c) ex Yorkshire Main Colliery, 15/7/1979;
     to Ashington Central Workshops, Northumberland, 28/10/1980; returned, 2/4/1981
(d) ex Kilnhurst Colliery, 5/1988
(e) ex Manvers Training Centre, after 5/1988, by 10/1988
(f) ex CE, Hatton, Derbyshire, /1988 (earlier Thoresby Colliery, Nottinghamshire)

(1) to Yorkshire Mining Museum, Caphouse Colliery, West Yorkshire, /1993 (by 6/1993)
(2) to Thoresby Colliery, Nottinghamshire, /1989

BILLINGLEY COLLIERY, Billingley

National Coal Board

Opened by NE2 by 1952.   CLOSED 7/1956

This drift mine had no standard gauge rail connection and locomotives were not used underground.

BIRDWELL CENTRAL WORKSHOPS, Birdwell

National Coal Board

Opened by NE5 5/1952; BNY from 26/3/1967; HQ 6/1967.   CLOSED after 1968

The workshops were served by a siding (opened 7/1955) on the north west side of the BR (ex LNER) line, ½ mile north east of Birdwell & Hoyland Common Station. Prior to 1955 locos arrived and left by road.

Gauge : 4ft 8½in

PRIOR	0-4-0ST	OC	K	3881	1899	(a)	(1)
TCD	0-4-0ST	OC	YE	1891	1923	(b)	(2)
VENTURE	0-4-0DM		JF	22287	1938	(c)	(3)
-	0-6-0ST	OC	YE	1889	1923	(d)	(4)
SENTINEL No.2	4wVBT	VCG	S	9401	1950	(e)	(5)
SENTINEL No.5	4wVBT	VCG	S	9570	1954	(f)	(6)

SENTINEL No.4	4wVBT	VCG	S	9557	1953	(g)	(7)
(75131) 3181 AUSTERITY	0-6-0ST	IC	HE	3181	1944	(h)	(8)
SENTINEL No.6	4wVBT	VCG	S	9616	1957	(j)	(9)
YORK No.2	0-4-0ST	OC	YE	2473	1949	(k)	(10)
YORK No.1	0-4-0ST	OC	YE	2474	1949	(m)	(11)

(a)   ex Monk Bretton Colliery, after 4/1953, by 7/1953
(b)   ex Smithywood Coking Plant, c/1954 (by 15/5/1954)
(c)   ex Barrow Colliery, after 5/1955, by 9/1955
         to Grange Colliery, after 29/4/1956, by 3/1957; returned after 9/1955 by 3/1957
         to Skiers Spring Colliery, after 7/1958, by 5/1959; returned c/1960
(d)   ex Barnsley Main Colliery, after 4/1957, by 7/1958
(e)   ex Smithy Wood Colliery, after 4/1958
(f)   ex Smithy Wood Colliery, c/1960 (by 3/1960)
(g)   ex Skiers Spring Colliery, after 8/1960, by 3/1961;
         to Wharncliffe Silkstone Colliery, c12/1960 (by 3/1961); returned, /1965
(h)   ex Barrow Colliery, after 1/1962, by 4/1962
(j)   ex Smithy Wood Colliery, 13/4/1961;
         to Smithy Wood Colliery after 11/1961, by 4/1962; returned, 10/1/1963
(k)   ex Barrow Colliery, c/1960 (by 12/3/1961)
(m)   ex Monk Bretton Colliery, after 6/1963, by 3/1964

(1)   to Monk Bretton Colliery, 26/8/1955
(2)   to Smithywood Coking Plant after 5/1954, by 16/4/1955
(3)   to Thos.W.Ward Ltd, Sheffield. 5/1961
(4)   to Barnsley Main Colliery, after 7/1958, by 6/1959
(5)   to Wombwell Colliery, 10/1959
(6)   to Smithy Wood Colliery, 7/3/1961
(7)   dismantled 5/1965 and frame to TH, Kilnhurst, 3/6/1965
(8)   to Barrow Colliery, 19/3/1964
(9)   to Smithy Wood Colliery, 3/5/1963
(10) to Barrow Colliery, after 14/10/1961, by 3/1964
(11) to Monk Bretton Colliery, after 5/1965, by 2/1966

BIRDWELL DISPOSAL POINT, Birdwell   N11

Used by MFP in 1946 only   SE 346008 ?

A rail loading point was located to the west of Rockingham Colliery on the LMSR Pilley branch. A photograph shows the locomotive stabled under an overline bridge, possibly at SE 346008 (Sheffield Road)

Gauge : 4ft 8½in

No.6	THORNCLIFFE	0-6-0ST	IC	MW	541	1875	(a)	(1)

(a)   ex Newton Chambers & Co Ltd, Thorncliffe Ironworks, hire, c/1946

(1)   returned to Newton Chambers & Co Ltd, Thorncliffe Ironworks, off hire, c/1946

BIRLEY EAST TRAINING CENTRE, Woodhouse   N12
ex **Sheffield Coal Co Ltd**   SK 423841

NE1 from 1/1/1947.   Training Centre CLOSED 1948; Railway CLOSED 1950.

This closed colliery, converted to a Training Centre in 1944, was served by sidings on the south side of the BR (ex LNER) owned, but NCB worked, Birley Colliery branch, 1 mile west of its junction with the main line at Woodhouse East Junction (¼ mile south-east of Woodhouse Station). Whilst very small amounts of coal were produced at the Training Centre, the main purpose of the locomotives kept in the shed there was to serve BIRLEY WEST LANDSALE DEPOT (SK 400845). This was located on the site of the closed colliery of that name 1½ miles west of the training centre and closed in 1950.

Reference : "'Winding Up', A History of Birley East Colliery", Alan Rowles, Author, 1992.

Gauge : 4ft 8½in

ORIENT		0-6-0ST	IC	HC	365	1890	(a)	(1)

(a)   ex Sheffield Coal Co Ltd, with site, 1/1/1947

(1)   to Brookhouse Colliery, /1950 (after 24/9/1950)

BIRLEY WEST LANDSALE DEPOT, Intake   N13
ex **Sheffield Coal Co Ltd**   SK 400845

NE1 from 1/1/1947.   CLOSED 1950

see BIRLEY EAST TRAINING CENTRE

BRIERLEY COLLIERY, Brierley   N14
ex **Hodroyd Coal Co Ltd**   SE 410109

NE4 from 1/1/1947.   CLOSED 1/1947

The colliery was served by a narrow gauge tubway which ran north and then north-east for 2 miles from FERRYMOOR COLLIERY (where the screens were located). No known locomotives. The colliery became BRIERLEY TRAINING CENTRE after closure.

BRODSWORTH COLLIERY, Adwick le Street   N15
ex **Doncaster Amalgamated Collieries Ltd**   SE 528077

NE2 from 1/1/1947; DCR from 26/3/1967; SYK from 1/10/1985; SYG from 1/4/1990. CLOSED 7/9/1990

The colliery was served by sidings at the end of a BR branch which ran west for 2½ miles from Castle Hill Junction on the Doncaster - Wakefield line. The colliery sidings were also connected at their west end to BR at Pickburn Station until the line there closed. There was an extensive 3ft 0in gauge surface system which replaced an earlier one of 2ft 0in gauge. Locomotives were used for all purposes underground as the gauge was progressively changed. Use of standard gauge locomotives ceased from 1988 after which all remaining traffic was rapid loaded.

Gauge : 4ft 8½in

No.7		0-6-0ST	OC	P	1117	1907	(a)	(1)
No.26	JOHN	0-6-0ST	OC	AE	1949	1924	(a)	(2)
(No.34	CLEMENT)	0-6-0ST	IC	HE	1983	1940	(b)	(3)
(No.30)	BR/M/22	0-6-0DM		HE	1724	1934		
		reb to 0-6-0DH		HE		1962	(c)	(4)
No.43	BRODSWORTH	BRM 21 3219/007						
		0-6-0DM		HE	4513	1955	New	(5)

	BULLCROFT No.2	0-6-0ST	OC	YE	1787	1922	(d)	(6)
D2613	BRM 5481 3219/008	0-6-0DM		HE	5662	1960	(e)	(7)
D2511	BRM 5477	0-6-0DM		HC	D1202	1961	(f)	(8)
No.47	BR/M226 3219/011	0-6-0DM		HC	D1068	1958	(g)	(9)
3	BR/M/6499	0-6-0DM		HC	D1189	1959		
		reb		HE	8901	1978	(h)	(10)

(a)   ex Doncaster Amalgamated Collieries Ltd, with site, 1/1/1947
(b)   ex Doncaster Amalgamated Collieries Ltd, with site, 1/1/1947;
       to Yorkshire Main Colliery after 5/1963, by 7/1963; returned after 7/1963, by 12/1963;
       to Hickleton Colliery, 4/1964; returned, 29/5/1964
(c)   ex HE, Leeds, West Yorkshire, after 4/1953, by 7/1955 (originally LMSR 7054)
       to HE c/1961; returned after 5/1961 by 7/1962;
       to Hickleton Colliery after 1/1969, by 27/4/1969; ex Bullcroft Colliery after 8/1970, by 5/1971
(d)   ex Bullcroft Colliery, 9/1961 (by 19/9/1961)
(e)   ex BR, Goole, East Yorkshire, 5/1968
(f)   ex BR, Barrow in Furness, Cumbria, 5/1968
(g)   ex Bullcroft Colliery, after 5/1971, by 3/1972
(h)   ex HE, Leeds, West Yorkshire, /1981

(1)   dismantled, 5/1963; scrapped, by 7/1964
(2)   scrapped, after 4/1957, by 4/1958
(3)   dismantled by 10/1968; boiler & tank scrapped by 8/1970;
       remainder scrapped after 4/1972, by 4/1973
(4)   to Hickleton Colliery, after 4/1972, by 3/1973
(5)   scrapped, after 11/1980, by 11/1981
(6)   to Carcroft Central Workshops, by 7/1962
(7)   to Bentley Colliery, after 12/12/1973, by 8/1975
(8)   to Keighley & Worth Valley Railway, Haworth, West Yorkshire, 8/10/1977
(9)   scrapped on site by C.F. Booth Ltd, after 2/5/1986, by 26/5/1987
(10) to Booth Roe Metals Ltd, Rotherham, 13/2/1990 and scrapped there, 2/1990

Gauge : 3ft 0in (Surface Stockyard)

BRM301	390/BR/M/301	4wDMF	RH	249550	1947	(a)	(1)
	390/BR/M/302 JOHN	4wDMF	RH	249552	1947	(a)	Scr c/1977
	390/BR/M/303	4wDMF	RH	249554	1947	(a)	(2)
No.0	390/BR/M/308	0-4-0DMF	HE	3426	1946	(b)	(3)

(a)   ex underground, c/1959 (by 5/1960)
(b)   ex underground, 11/11/1974

(1)   sold or scrapped after 22/6/1986, by 6/6/1987
(2)   dismantled, /1964; scrapped 3/1965 or 4/1965
(3)   sold or scrapped, after 19/2/1984, by 14/9/1985

Gauge : 3ft 0in (Underground locomotives)

No.	390/BR/M/308	0-4-0DMF	HE	3426	1946	(a)	(1)
	390/BR/M/301	4wDMF	RH	249550	1947	New	(2)
	390/BR/M/302 JOHN	4wDMF	RH	249552	1947	New	(2)
	390/BR/M/303	4wDMF	RH	249554	1947	(b)	(2)
	390/BR/M/304 JOE	0-4-0DMF	HE	3611	1948	New	(3)
No.5	390/BR/M/305 HAROLD	0-4-0DMF	HE	3612	1948	New	(3)
No.6	390/BR/M/306 CECIL	0-4-0DMF	HE	3613	1948	New	(3)
	390/H/M/438	0-4-0DMF	HE	3287	1945	(c)	(4)

	390/BR/M/307	0-4-0DMF	HE	3618	1948	(d)	(5)
No.1	390/BR/M/311	0-6-0DMF	HC	DM857	1954	New	(3)
No.4	390/BR/M/314	0-6-0DMF	HC	DM858	1954	New	(6)
No.3	390/BR/M/310	0-6-0DMF	HC	DM859	1954	New	(3)
		0-6-0DMF	HC	DM860	1954	New	(7)
No.8	390/BR/M/312	0-6-0DMF	HC	DM903	1956	New	(3)
	390/BR/M/313	0-6-0DMF	HC	DM904	1956	New	(8)
No.5	390/BR/M/315	0-6-0DMF	HC	DM905	1956	New	(3)
No.6	390/BR/M/316	0-6-0DMF	HC	DM906	1956	New	(3)
No.7	390/BR/M/317	0-6-0DMF	HC	DM907	1956	New	(3)
	390/BR/M/318	0-6-0DMF	HC	DM908	1956	New	(9)
No.16	390/BR/M/319	0-6-0DMF	HC	DM909	1956	New	(3)
No.14	390/BR/M/320	0-6-0DMF	HC	DM910	1957	New	(3)
No.3	390/BR/M/2646	0-6-0DMF	HC	DM911	1957	New	(3)
No.9	390/BR/M/2797	0-6-0DMF	HC	DM912	1957	New	(3)
	-	0-6-0DMF	HC	DM913	1958	New	(10)
No.10	390/BR/M/775	0-6-0DMF	HC	DM914	1958	New	(11)
No.11	390/BR/M/830	0-6-0DMF	HC	DM1110	1959	New	(3)
No.1	390/BR/M/774	0-6-0DMF	HC	DM1120	1958	(e)	(12)
No.8	390/BR/M/897	0-6-0DMF	HC	DM1213	1960	New	(3)
No.12	390/BR/M/898	0-6-0DMF	HC	DM1214	1960	New	(3)
No.13	390/BR/M/3478	0-6-0DMF	HC	DM1215	1960	New	(3)
No.14	390/BR/M/	0-6-0DMF	HC	DM1284	1962	New	s/s by 9/1979
No.15	390/BR/M/	0-6-0DMF	HC	DM1285	1962	New	(3)
No.8	141/1988	4wBEF	CE	B1575F	1978	(f)	(13)
	-	4wBEF	CE	B2959B	1982	(f)	(13)

(a)   ex Doncaster Amalgamated Collieries Ltd, with site, 1/1/1947
(b)   ex Yorkshire Main Colliery, 7/1948
(c)   ex Yorkshire Main Colliery, by /1959
(d)   ex Yorkshire Main Colliery, 11/1954
(e)   ex Yorkshire Main Colliery, 11/1958
(f)   ex CE, Hatton, Derbyshire, /1989; earlier at Manton Colliery, Nottinghamshire

(1)   to surface, 11/11/1974
(2)   to surface c/1959 (by 5/1960)
(3)   sold, scrapped or abandoned underground, by /1991
(4)   to Hickleton Colliery, by 6/1959
(5)   written off, c/1973
(6)   dismantled on surface, 11/1981; scrapped, c/1982
(7)   to Yorkshire Main Colliery, 17/10/1955
(8)   to Hickleton Colliery, by 10/1975
(9)   to Hickleton Colliery, 10/11/1964
(10) to HC, Leeds, West Yorkshire, 6/1958; thence to Yorkshire Main Colliery
(11) to Bentley Training Centre, 9/1/1979
(12) to Hickleton Colliery, 6/1968
(13) to Silverwood Colliery, /1990

Gauge : 2ft 0in (surface stockyard)

	-	4wDM	HE	1935	1938	(a)	(1)
	-	4wDM	RH	211644	1941	(b)	(2)

(a)   ex Doncaster Amalgamated Collieries Ltd, with site, 1/1/1947

(b) ex George W. Bungey Ltd, dealers, Hayes, Middlesex, c/1949;
     originally Ministry of Supply

(1) out of use by /1961; scrapped after 4/1968, by 10/1968
(2) to Bullcroft Colliery, after 7/1960, by 7/1961

Gauge : 200mm (Roadrailer trapped rail system - surface demonstration only)
         (note that in this list '1aD' indicates '1 axle driven'.)

No.3   1adDHF   BGB   DRL25/1/202 1970 (a)   (1)

(a) ex Desford Colliery, Leicestershire, 3/1970

(1) to Desford Colliery, Leicestershire, 3/1970

BROOKHOUSE COLLIERY, Beighton   N16
ex Sheffield Coal Co Ltd   SK 454843
NE1 from 1/1/1947: SYK from 26/3/1967.   Rail traffic ceased by 9/1982; CLOSED 10/1985
Served by sidings on the south side of the BR (ex LNER) line, ¾ miles west of Waleswood Station. A
line ran west at a lower level from the colliery for ½ mile to the works of United Coke & Chemicals Ltd
(SK 449843) where the shunting was carried out by the NCB until c1952.

Gauge : 4ft 8½in

	BIRLEY No.5	0-4-0ST	OC	P	1454	1917	(a)	(1)
	VICTORY	0-4-0ST	OC	AB	1654	1920	(b)	(2)
	BIRLEY No.6	0-4-0ST	OC	P	1653	1925	(b)	(3)
TRG	(WDG until after 1961)	0-6-0ST	IC	P	1634	1927	(b)	Scr 10/1966
WDG	(TRG until after 1961)	0-6-0ST	IC	P	1765	1929	(b)	Scr 10/1966
No.10	HUNTSMAN	0-6-0ST	OC	AB	2018	1936	(c)	(4)
	ORIENT	0-6-0ST	IC	HC	365	1890	(d)	(5)
No.7		0-6-0ST	IC	HL	3726	1928	(e)	(6)
		0-6-0DH		YE	2913	1965	New	(7)
DL4	No.4	0-6-0DM		HC	D1152	1959	(f)	(8)
D2229	No.5 521/1952	0-6-0DM		VF	D278	1955		
				DC	2552	1955	(g)	(9)

(a) ex Sheffield Coal Co Ltd, with site, 1/1/1947;
     to Maltby Colliery, /1948; returned, after 18/4/1949, by 19/5/1951
(b) ex Sheffield Coal Co Ltd, with site, 1/1/1947;
     to Maltby Colliery, /1948; returned, after 18/4/1949, by 19/5/1951
(c) ex Treeton Colliery, /1949 (by 18/4/1949);
     to Handsworth Colliery, after 18/4/1949, by 1/1/1950; ex Treeton Colliery, 4/1954 or 5/1954
(d) ex Birley East Training Centre, /1950 (after 24/9/1950)
(e) ex Nunnery Colliery, 2/1954
(f) ex Treeton Colliery, 13/12/1969
(g) ex BR, Thornaby, Cleveland, 28/8/1970;
     to Orgreave Colliery, /1971 (by 28/7/1971); returned c1/1972 (by 19/4/1972);
     to Orgreave Colliery, after 5/1973, by 11/1973; returned after 11/1973, by 7/1974

(1) to Maltby Colliery after 6/1951, by 6/7/1952
(2) to Maltby Colliery, after 18/4/1949, by 4/1950
(3) scrapped on site by C.F. Booth Ltd, of Rotherham, after 10/1967, by 2/1968
(4) to Handsworth Colliery, after 9/1955, by 12/1955
(5) scrapped on site by Marple & Gillott Ltd, of Sheffield, 6/1960
(6) scrapped on site, 8/1972
(7) to Treeton Colliery, c/1969 (after 9/1968, by 6/1972)

(8)   to Manton Colliery, Nottinghamshire, 29/3/1983
(9)   to Manton Colliery, Nottinghamshire, 28/3/1983

BULLCROFT COLLIERY, Carcroft   N17
ex **Doncaster Amalgamated Collieries Ltd**   SE 541097
NE2 from 1/1/1947; DCR from 26/3/1967.   Merged with BRODSWORTH COLLIERY 9/1970
Rail traffic ceased 1971

The colliery was served by sidings on the north side of the BR (ex LNER) line from Adwick Junction to Stainforth & Hatfield Station, ½ mile east of Adwick Junction. It was, until 20/10/1958, also served by a BR branch from Bullcroft Junction on the former H&B and GC Joint line. There was a connection to CARCROFT CENTRAL WORKSHOPS, located at the east end of the colliery yard and a line which ran east to dirt tips (1 mile). The underground locomotives were used for coal haulage until coal winding was transferred to Brodsworth in 1971

Gauge : 4ft 8½in

In addition to the locomotives listed below, various 200HP 0-6-0DM locomotives were hired from British Railways in 1969.

	(THE) BULL	0-4-0ST	OC	Mkm		1909	(a)	(1)
	BEATTY	0-4-0ST	OC	MW	1902	1916	(a)	(1)
No.2	TOM	0-6-0ST	IC	HE	1826	1939	(b)	(2)
	BULLCROFT No.2	0-6-0ST	OC	YE	1787	1922	(c)	(3)
No.47	BU/M226 3219/011	0-6-0DM		HC	D1068	1958	New (d)	(4)
No.50		0-6-0ST	OC	HC	1532	1924	(e)	(5)
No.30		0-6-0DM		HE	1724	1934		
	Rebuilt to 0-6-0DH			HE		1962	(f)	(6)

(a)   ex Doncaster Amalgamated Collieries Ltd, with site, 1/1/1947
(b)   ex Doncaster Amalgamated Collieries Ltd, with site, 1/1/1947;
       to Markham Main Colliery, after 7/1962, by 2/1963; returned, after 2/1963, by 7/1963;
       to Markham Main Colliery, after 5/1965; returned, after 7/1967, by 6/1968
(c)   ex Hatfield Colliery, 3/1957
(d)   to Allerton Bywater Central Workshops, West Yorkshire, after 20/3/1969, by 30/3/1969;
       returned after 12/1969, by 1/1970
(e)   ex Rossington Colliery, after 6/1967, by 6/1968
(f)   ex Hickleton Colliery, after 6/1969, by 11/1969

(1)   scrapped by Thos.W.Ward Ltd, 4/1958
(2)   parts to Askern Colliery, 3/1969. Remains scrapped on site by George Cohen, Sons & Co Ltd, Tinsley, Sheffield, 11/1969
(3)   to Brodsworth Colliery, /1961 (by 19/9/1961)
(4)   to Brodsworth Colliery after 5/1971, by 3/1972
(5)   scrapped by Raynor Contracting Ltd, Sheffield, 11/1969 (after 20/11/1969)
(6)   to Brodsworth Colliery, after 8/1970, by 5/1971

Gauge : 2ft 0in (Surface stockyard)

	BU/M/225	4wDM	RH	211644	1941	(a)	(1)
	ALBERT	4wDM	RH	223748	1946	(b)	(1)

(a)   ex Brodsworth Colliery after 7/1960, by 7/1961
(b)   origin unknown; here by /1970 (originally Ministry of Supply)

(1)   sold or scrapped c/1972 (by 1974)

Gauge : 3ft 0in (Underground locomotives)

-	0-6-0DMF	HC	DM721	1951	New	(1)
-	0-6-0DMF	HC	DM722	1951	New	(1)
-	0-6-0DMF	HC	DM723	1951	New	(1)

(1)   written off c/1971

CADEBY COLLIERY, Conisbrough   N18
ex **Amalgamated Denaby Collieries Ltd**   SE 512997
NE3 from 1/1/1947; SYK from 26/3/1967.   CLOSED 11/1986

The colliery was located north of the BR Mexborough-Doncaster line to which siding connection was made east of Conisbrough Station. There was a second connection to the Wrangbrook Junction line, north east of Lowfield Junction. A third main line connection with the former LMSR Dearne Valley Railway north of the colliery had closed by 1958. Connection was made, via the BR bridge over the River Don on the Lowfield Junction line, with Denaby Colliery, 1 mile to the west. There had always been interworking between these collieries. Coal went to a staith at Denaby and Cadeby dirt was taken by rail to a tip near there until 1964. Cadeby locomotives were repaired at Denaby and spare locomotives were kept there, with Cadeby only having the working stock. With closure of the remaining part of the Denaby surface in the 1960s, railway operations were concentrated at Cadeby. We list here all locomotives that have worked at either location, notwithstanding that the majority were at Denaby in the 1950s. From 1959, after which Denaby Coal was wound at Cadeby, the two systems were closely integrated. From 1950 locomotives were used underground on a 3ft 0in gauge line to haul coal to the shaft and the supplies tracks were later altered to that gauge.

Gauge : 4ft 8½in

5	CLIFTON	0-6-0ST	OC	BH	1024	1888	(a)	(1)
7	CONISBRO	0-6-0ST	IC	BP	3872	1896	(a)	(2)
(No.12	MELTON)	0-6-0ST	IC	P	824	1900		
		reb		HC		1927	(a)	(3)
No.12A	FIRSBY	0-6-0ST	IC	P	836	1901		
		reb		HC		1926	(a)	
		reb with parts of		P	824	1962		(4)
No.43	(ABERCONWAY)							
	JENKY	0-6-0ST	IC	RSHN	6942	1938	(a)	(5)
No.16	THRYBERGH	0-6-0ST	OC	HC	692	1904	(a)	(6)
18	DENABY	0-6-0ST	OC	HC	649	1903	(a)	(7)
	RERESBY	0-6-0ST	IC	HC	704	1904		
		reb				1926	(a)	(8)
No.20	CADEBY	0-6-0ST	IC	HC	920	1910	(a)	(9)
No.38	(SPROTBOROUGH)	0-6-0ST	IC	HL	3658	1926	(b)	Scr 7/1966
26		0-6-0ST	OC	HC	1629	1928	(c)	Scr /1962
No.8	DICK No.55	0-6-0DM		HC	D1115	1958	New	(10)
No.7	(CHARLIE No.62)	0-6-0DM		HC	D1139	1959	New	(11)
No.17		0-4-0ST	OC	HC	751	1906	(d)	Scr /1962
64		0-6-0ST	IC	RSHN	7204	1944		
		reb		HE	3884	1963	(e)	(12)
No.10	ATLAS No.15	0-4-0ST	OC	HL	2464	1900	(f)	(13)
	KEN No.67 521/5331	0-6-0DH		S	10180	1964	New (g)	(14)
66		0-6-0ST	IC	HE	3890	1964	New	(15)
	FRANK No.70	0-6-0DH		RR	10223	1965	New	(16)
D2513		0-6-0DM		HC	D1204	1961	(h)	(17)

D2208		0-6-0DM		VF	D209	1953		
				DC	2483	1953	(j)	(18)
No.35		0-6-0ST	OC	AB	1792	1923		
			reb	AB		1938	(k)	(19)
No.20		0-6-0T	OC	HC	1731	1942	(m)	(20)
D2332)	LLOYD	0-6-0DM		RSHN	8191	1961		
				DC	2713	1961	(n)	(21)
No.22	DAVID No.58	0-6-0DM		HC	D1128	1958	(p)	(10)

(a)   ex Amalgamated Denaby Collieries Ltd, with site, 1/1/1947
(b)   ex Amalgamated Denaby Collieries Ltd, with site, 1/1/1947;
      to Elsecar Central Workshops after 5/1959, by 3/1961; returned after 10/1961, by 7/1962
(c)   ex Eagre Construction Co Ltd, Scunthorpe, Lincolnshire, /1957
(d)   ex Aldwarke Main Colliery, 9/1961
(e)   ex HE, Leeds, West Yorkshire, 7/1963;
      earlier at WD, Bicester Depot, Oxfordshire, WD 176 (ex 75274)
(f)   ex Silverwood Colliery, 13/8/1963
(g)   to HE, Leeds, West Yorkshire, 14/6/1975, but transferred to TH, Kilnhurst, 12/1975;
      returned from TH, Kilnhurst, 14/6/1976
(h)   ex BR, Barrow in Furness, Cumbria, 12/1968
(j)   ex Cortonwood Colliery, after 3/1959, by 5/1959
(k)   ex Cortonwood Colliery, after 3/1969, by 19/4/1970
(m) ex Orgreave Colliery, 3/1970
(n)   ex Manvers Colliery, 28/8/1975
(p)   ex New Stubbin Colliery, 23/8/1978

(1)   scrapped on site by T.H. Briggs Ltd, c/1961
(2)   scrapped, after 3/1964, by 11/1964
(3)   parts used to rebuild P 836 and remains scrapped at Denaby, /1963
(4)   scrapped, after 2/1973, by 4/1973
(5)   to Elsecar Central Workshops, after 10/1962, by 2/1963
(6)   scrapped, after 9/1958, by 12/3/1961
(7)   scrapped, after 8/1958, by 2/1959
(8)   to Rossington Colliery, c/1948
(9)   dismantled, 8/1964; scrapped by 12/1964
(10) scrapped on site by C. & E. Thornton Ltd, of Sheffield, after 27/4/1987, by 7/5/1987
(11) out of use by /1976; scrapped on site by Ernest Northcliffe Ltd, of Parkgate, 10/1980
(12) scrapped, after 3/1970, by 6/1970
(13) to Silverwood Colliery, 19/9/1963
(14) to Maltby Colliery, 8/5/1987
(15) to Quainton Railway Society, Quainton Road, Buckinghamshire,
      after 20/10/1975, by 4/11/1975
(16) to Manvers Colliery, after 7/1971, by 9/1971
(17) scrapped on site, 10/1975
(18) to Silverwood Colliery, after 8/1970, by 28/7/1971
(19) scrapped on site 4/1973
(20) to North Yorkshire Moors Railway, North Yorkshire, after 8/1972, by 2/1973
(21) to Thurcroft Colliery, 14/6/1976

Gauge : 3ft 0in (Surface stockyard)

-	4wDM	MR	9695	1952	(a)	(1)
7143	4wDM	MR	7406	1939	(b)	s/s c/1972
-	4wDM	MR	8814	1943	(c)	(2)
-	4wDM	MR	9696	1952	(d)	s/s c/1972

No.10	IVOR	390/6014	0-6-0DMF	HE	3517	1948	(e)	(3)
	KATY	521/13002	4wDM	MR	40S280	1968	(f)	(4)

(a)   ex Manvers Colliery, c/1958 (by 9/1958) (uncertain if regauged from 2ft 0in or 2ft 2in)
(b)   regauged from 1ft 11in gauge at unknown date (by 1/1964) – see 2ft 0in gauge list below
(c)   regauged from 1ft 11in gauge at unknown date (by 11/1968)
(d)   ex Kilnhurst Colliery, after 12/1968, by 8/1970
(e)   ex underground, /1979
(f)   ex Silverwood Colliery, 2/12/1981

(1)   to Wath Colliery (where it was 2ft 2in gauge), 11/1959
(2)   scrapped, /1980 (after 4/1980)
(3)   to Silverwood Colliery, after 22/2/1987, by 2/4/1987
(4)   to Kiveton Park Colliery, 15/12/1986

Gauge : 3ft 0in (Underground locomotives)

No.7		390/6012		0-6-0DMF	HE	3514	1948	New	(1)
No.8				0-6-0DMF	HE	3515	1948	New	(2)
No.9	No.1	390/6013		0-6-0DMF	HE	3516	1948	New	(3)
No.10		390/6014		0-6-0DMF	HE	3517	1948	New (a)	(4)
No.16		390/6015		0-6-0DMF	HE	4035	1949	New	(1)
No.35	No.5	390/6018		0-6-0DMF	HE	4862	1956	(b)	(1)
No.42	No.3	390/6010		0-6-0DMF	HE	4863	1956	(c)	(5)
No.36		390/6019		0-6-0DMF	HE	4864	1956	(d)	(1)
No.37		390/6017		0-6-0DMF	HE	4865	1957	New	(1)
No.38		390/6016		0-6-0DMF	HE	5209	1957	New	(1)
No.39		390/6011		0-6-0DMF	HE	5210	1957	New (e)	(1)
No.40		390/6007		0-6-0DMF	HE	5211	1957	New	(1)
No.41	No.52	PR6011	390/6008						
				0-6-0DMF	HE	5214	1957	New (f)	(1)
No.50		390/6005		0-6-0DMF	HE	5430	1958	New	(1)
No.51		390/6006		0-6-0DMF	HE	5431	1958	New	(1)
No.56	No.34	390/6009		0-6-0DMF	HE	3432	1948	(g)	(1)
No.58				0-6-0DMF	HE	4054	1952	(h)	(6)
No.15				0-6-0DMF	HE	4032	1949	(h)	(7)

(a)   to Ashington Central Workshops, Northumberland, 22/11/1976; returned, 18/7/1978
(b)   ex Silverwood Colliery, 11/1956
(c)   ex Kilnhurst Colliery, 13/9/1956
(d)   ex Kilnhurst Colliery, 11/6/1957
(e)   to Ashington Central Workshops, Northumberland, 13/8/1980; returned, 24/9/1981
(f)   to Silverwood Colliery; returned;
        to Ashington Central Workshops, Northumberland,16/10/1980; returned, 21/12/1981
(g)   ex Upton Colliery, West Yorkshire, by 11/10/1962
(h)   ex Silverwood Colliery, 4/1981

(1)   abandoned underground 11/1986
(2)   to Manvers Colliery, 6/1954
(3)   to Kilnhurst Colliery, c/1972
(4)   to surface stockyard, /1979
(5)   to Kilnhurst Colliery, by /1969
(6)   dismantled for spares and remains abandoned underground, /1986
(7)   dismantled for spares and remains scrapped. c/1982

Gauge : 2ft 0in (surface stockyard)

		4wDM	MR	7406	1939	(a)	(1)
-							

(a)   ex unknown location, by 3/1957 (new to Mark Williams & Co Ltd, Cheltenham, Gloucestershire, per Diesel Loco Hirers Ltd.)

(1)   rebuilt to 3ft 0in gauge at unknown date (by 1/1964)

Gauge : 1ft 11in (Surface locomotives, not used here on this gauge)

		4wDM	MR	8814	1943	(a)	(1)
-		4wDM	MR	7606	1939	(b)	(2)

(a)   ex Denaby Colliery, c/1965 (by 10/1965)
(b)   ex Denaby Colliery, c/1966 (by 11/1968)

(1)   rebuilt to 3ft 0in gauge at unknown date (by 11/1968)
(2)   dismantled for spares c/1968; remains sold or scrapped c/1972

Gauge : 1ft 11in (Underground locomotives, not used here on this gauge) -
These locomotives were moved here underground from Denaby Colliery and left near Cadeby pit bottom.

			4wDMF	RH	268860	1949	(a)	(1)
No.1	No.55	390/10207	0-4-0DMF	HE	4129	1950	(a)	(1)
No.2			0-4-0DMF	HE	4130	1950	(a)	(2)
No.3			0-4-0DMF	HE	4332	1950	(a)	(3)
No.4	No.41	390/10210	0-4-0DMF	HE	4333	1950	(a)	(1)
No.5	No.48		0-4-0DMF	HE	4808	1954	(a)	(4)
6			0-4-0DMF	HE	5203	1957	(a)	(3)

(a)   ex Denaby Colliery, /1968 (by 11/1968)

(1)   abandoned underground near Cadeby shafts, c/1969; still accessible, 2/1981
(2)   to HE, Leeds, West Yorkshire, c/1970 (after 6/1969);
       regauged to 2ft 2in and thence to Barnburgh Colliery
(3)   HE 4332 was out of use on surface 11/1968 and 6/1969 – then -
       NCB records show HE 5203 as moving to Hunslet in 1970, repaired as HE 7223 and
       regauged to 2ft 2in for Barnburgh Colliery. But a locomotive carrying HE 5203 plates and
       correctly numbered 6 was out of use on Cadeby surface until at least 8/1975, after which it
       was sold or scrapped. It is likely that HE 4332 was the locomotive which went to Hunslet in
       1970
(4)   to Kiveton Park Colliery c/1973 (after 6/1969)

CANKLOW SCREENS DISPOSAL POINT

Some sources use this title as a reference to ROTHERHAM MAIN DISPOSAL POINT, which see.

CARCROFT CENTRAL WORKSHOPS, Carcroft

National Coal Board

N19
SE 545096
CLOSED

Opened by NE2 1959; DCR from 26/3/1967; HQ from 6/1967.

Located at the east end of BULLCROFT COLLIERY sidings, to which it was connected.

Gauge : 4ft 8½in

	BENTLEY No.3	0-6-0T	IC	HC	850	1909	(a)	
	rebuilt Carcroft with parts of HC 851 & 925 from					1962		(1)
	BULLCROFT No.2	0-6-0ST	OC	YE	1787	1922	(b)	(2)
(75023)		0-6-0ST	IC	HE	2872	1943	(c)	(3)
	THORNE No.2	0-6-0ST	IC	HE	3804	1953	(d)	(4)
No.3		0-6-0ST	IC	P	1219	1910	(e)	(5)

(a)   ex Bentley Colliery, BENTLEY, after 7/1960, by 5/1961
(b)   ex Brodsworth Colliery, after 9/1961, by 7/1962
(c)   ex Bentley Colliery, /1963 (by 7/1963)
(d)   ex Thorne Colliery, c5/1963 (after 9/1961, by 7/1963)
(e)   ex Thorne Colliery, after 6/1963, by 3/1964

(1)   to Bentley Colliery, BENTLEY No.3, after [1?]/1962, by 7/1963
(2)   to Thorne Colliery, c6/1963 (after 7/1962, by 6/1963)
(3)   to Bentley Colliery, after 6/1964, by 3/1965
(4)   to Thorne Colliery, c10/1963 (by 3/1964)
(5)   to Thorne Colliery, after 5/1965, by 4/1966

Gauge : 3ft 0in

390/YM/1668/M No.2	0-4-0DMF		HE	3614	1948	(a)	(1)

(a)   ex Yorkshire Main Colliery, c/1966

(1)   to Yorkshire Main Colliery, c/1967

Gauge : 2ft 3in

BENTLEY No.2 390/BE/M/450	0-4-0DMF		HE	2662	1941	(a)	(1)
No.3 390/BE/M/482	0-4-0DMF		HE	2663	1941	(b)	(2)

(a)   ex Bentley Colliery, 6/2/1964
(b)   ex Bentley Colliery, 29/9/1961

(1)   to Bentley Colliery, c/1965
(2)   to Bentley Colliery, 16/10/1964

Gauge : 2ft 2in

390/H/M/458	0-6-0DMF		HC	DM956	1955	(a)	(1)

(a)   ex Hickleton Colliery, c/1965

(1)   to Hickleton Colliery, 21/4/1966

Gauge : 2ft 0in

	390/T/M/3050	0-6-0DMF	HC	DM928	1955	(a)	(1)
No.17	390/R/M/2192	0-6-0DMF	HC	DM934	1956	(b)	(2)
	390/MM/M/2569	0-6-0DMF	HC	DM796	1952	(c)	(3)
No.16	390/R/M/2180	0-6-0DMF	HC	DM933	1956	(d)	(4)

No.14	390/R/M/2156		0-6-0DMF	HC	DM929	1955	(e)	(5)
	390/MM/M/2521		0-4-0DMF	HC	DM749	1949	(f)	(6)
No.23	390MM/M/2557	4	0-4-0DMF	HC	DM752	1949	(g)	(7)
	390/MM/M/2593		0-6-0DMF	HC	DM798	1953	(h)	(8)
No.14	390/HA/M/0717	No.111	0-6-0DMF	HC	DM717	1949	(j)	(9)
No.9	390/HA/M/0677	No.109	0-6-0DMF	HC	DM677	1949	(k)	(10)
	390/MM/M/2655		0-6-0DMF	HC	DM979	1956	(m)	(11)
No.10	390/HA/M/0675	No.107	0-6-0DMF	HC	DM675	1949	(n)	(12)
No.21	390/HA/M/0980	No.101	0-6-0DMF	HC	DM980	1955	(p)	(13)
No.5	390/HA/M/0630	No.101	0-6-0DMF	HC	DM630	1948	(q)	(14)

(a)   ex Thorne Colliery, 17/6/1959
(b)   ex Rossington Colliery, 30/8/1961
(c)   ex Markham Main Colliery, 2/1/1962
(d)   ex Rossington Colliery, 21/2/1962
(e)   ex Rossington Colliery, 11/10/1962
(f)   ex Markham Main Colliery, 5/12/1962
(g)   ex Markham Main Colliery, 16/12/1963
(h)   ex Markham Main Colliery, 7/2/1963;
       to HC, Leeds, West Yorkshire, repairs, 2/1963; returned, 4/1963
(j)   ex Hatfield Colliery, 10/4/1963
(k)   ex Hatfield Colliery, 22/7/1963
(m)   ex Markham Main Colliery, 12/1963
(n)   ex Hatfield Colliery, /1964
(p)   ex Hatfield Colliery, 6/10/1965
(q)   ex Hatfield Colliery, c/1966

(1)   to Markham Main Colliery, 4/1/1960
(2)   to Rossington Colliery, 24/1/1962
(3)   to Hatfield Colliery, after /1962, by /1966
(4)   to Rossington Colliery, 25/9/1962
(5)   to Rossington Colliery, 29/1/1963
(6)   to Askern Colliery, 7/12/1972
(7)   to Markham Main Colliery, 8/3/1966
(8)   to Markham Main Colliery, 26/4/1963
(9)   to Hatfield Colliery, 5/11/1963
(10) to Hatfield Colliery, 11/12/1963
(11) converted to 2ft 2in gauge; to Hickleton Colliery, 7/1964
(12) to Ackton Hall Colliery, West Yorkshire, c/1966
(13) to Hatfield Colliery, 29/12/1966
(14) to Manvers Training Centre c/1967-1968 (by 10/1968)

CORTONWOOD COKING PLANT, Brampton   N20
ex **Cortonwood Collieries Co Ltd**   SE ?
NE(C) from 1/1/1947.   CLOSED 1963
Located adjacent to CORTONWOOD COLLIERY - which see.

CORTONWOOD COLLIERY, Brampton   N21

ex **Cortonwood Collieries Co Ltd**   SE 406016

NE3 from 1/1/1947; SYK 26/3/1967.   CLOSED 10/1985

Served by sidings which ran south from the BR Elsecar Goods branch, 1 mile south-west of Elsecar Junction. There were rail served dirt tips to the south and traffic was worked for the adjacent CORTONWOOD COKING PLANT. Use of standard gauge locomotives ceased by 1982.

Gauge : 4ft 8½in

No.26	(36) (DORIS)		0-6-0ST	OC	AB	1498	1918	(a)	(1)
No.25	(LESLIE)		0-4-0ST	OC	HL	3307	1918	(b)	(2)
	PETER		0-4-0ST	OC	P	1650	1924	(b)	Scr /1949
No.30	(RICHARD No.12)		0-6-0ST	OC	AE	1819	1919	(c)	(3)
No.35	(HALLAMSHIRE)		0-6-0ST	OC	AB	1792	1923		
				reb	AB		1938	(d)	(4)
No.27			0-4-0ST	OC	HC	1338	1918	(e)	(5)
44			0-6-0ST	IC	P	1891	1940	(f)	(6)
53			0-6-0ST	IC	HE	3856	1956	New (g)	(7)
D2208			0-6-0DM		VF	D209	1953		
					DC	2483	1953	(h)	(8)
D2317	No.10		0-6-0DM		RSHD	8176	1960		
					DC	2698	1960	(j)	(9)
No.14	WILF No.72 521/11944	0-6-0DH		HE	6287	1965	(k)	(10)
No.31	(D2328) DINNINGTON No.2	0-6-0DM		RSHD	8187	1961		
					DC	2709	1961	(m)	(11)

(a)   ex Cortonwood Collieries Ltd, with site 1/1/1947;
        to Manvers Colliery, after 7/1961, by 11/1962; returned, 11/1962 or 12/1962
(b)   ex Cortonwood Collieries Ltd, with site 1/1/1947
(c)   ex Thos.W.Ward Ltd, /1948; earlier WD Shoeburyness, Essex, WD 71660,
        and may have carried BRAMLEY No.6 on arrival
(d)   ex Kilnhurst Colliery, c/1948 (by 10/1949);
        to Elsecar Central Workshops, after 4/1960, by 12/3/1961;
        returned after 16/7/1961, by 12/1962
(e)   ex Elsecar Central Workshops, /1955
(f)   ex Manvers Colliery, after 8/1960, by 12/3/1961
(g)   to Elsecar Central Workshops, 15/1/1963; returned, after 6/1965, by 10/1965
(h)   ex Manvers Colliery, after 12/1968, by 30/2/1969
(j)   ex Manvers Colliery, 5/1970 (by 16/5/1970);
        [possibly to New Stubbin Colliery 6/1976 or 7/1976; returned, /1976 (after 7/1976) ?]
(k)   ex Elsecar Colliery, 9/1981
(m)   ex Kiveton Park Colliery, 18/7/1985

(1)   scrapped on site by G.B. Housley Ltd, c28/8/1970
(2)   scrapped, after 8/1953, by 5/1955
(3)   to Elsecar Central Workshops, 3/1956
(4)   to Cadeby Colliery, after 3/1969, by 6/1970
(5)   to Elsecar Colliery, /1955
(6)   to Manvers Colliery, after 16/7/1961, by 11/1962
(7)   scrapped on site by George Cohen, Sons & Co Ltd, 8/1970
(8)   to Cadeby Colliery, after 3/1969, by 5/1969
(9)   scrapped on site by Wath Skip Hire Ltd, 7/1986
(10)   sold or scrapped, after 6/1986, by 29/11/1986
(11)   scrapped on site by Wath Skip Hire Ltd, after 6/1986, by 29/11/1986

DALTON MAIN COKING PLANT, Thrybergh

N22

ex **Dalton Main Collieries Ltd**

SK 478938

NE(C) from 1/1/1947.

CLOSED 1960

Adjacent to SILVERWOOD COLLIERY - which see.

Gauge : 4ft 8½in

-	0-4-0WE	GB	1348	1934	(a)	(1)

(a)   ex Dalton Main Collieries Ltd, with site, 1/1/1947

(1)   to Handsworth (High Hazels) Coking Plant, c/1954

DARFIELD COLLIERY, Wombwell

N23

ex **Mitchell Main Colliery Co Ltd**

SE 401039

NE4 from 1/1/1947; NE5 from 26/4/1964; BNY from 26/3/1967.
Merged with HOUGHTON MAIN 11/1986 and coal winding ceased.

Served by sidings which ran north from the BR (ex LNER) line, ½ mile north-west of Wombwell Station. Use of standard gauge locomotives ceased 1982. The trapped rail underground locomotives were used for man riding and supplies. The duties of the earlier ones are not known.

Gauge : 4ft 8½in

No.2		0-4-0ST	OC	HC	291	1887		
			reb	HC		1935	(a)	(1)
	DARFIELD	0-4-0ST	OC	P	992	1905	(a)	(2)
	-	0-4-0ST	OC	P	2108	1950	New	(3)
SUR28	DARFIELD No.1	0-6-0ST	IC	HE	3783	1953	New (b)	(4)
SUR29	DARFIELD No.2	0-6-0ST	IC	HE	3805	1953	New	(5)
TL10		4wDH		TH	171C	1966		
	built on frame of			S			New	(6)
	-	0-4-0DH		RR	10203	1964	(c)	(7)

(a)   ex Mitchell Main Colly Co Ltd, with site, 1/1/1947
(b)   to Houghton Main Colliery, 15/7/1955; returned, /1959
(c)   ex TH, Kilnhurst, /1974 (by 12/1974)

(1)   scrapped, after 9/9/1951, by 6/1952
(2)   to Mitchell Main Colliery, 6/5/1951
(3)   to Mitchell Main Colliery, 17/7/1954
(4)   to A. Hall, Delph, Oldham, Greater Manchester,11/1974 or 12/1974
(5)   parts recovered for HE 3783 and remains scrapped on site by C.E.Demolition Ltd, 12/1974
(6)   to Coopers (Metals) Ltd, Sheffield, after 28/10/1981, by 18/1/1982,
       and scrapped there c10/3/1982
(7)   to Smithywood Coking Plant, 20/6/1985

Gauge : 2ft 1in (Underground locomotives)

No.1	0-4-0DMF	HC	DM703	1947	(a)	(1)
-	4wDM	RH	249563	1947	(b)	Scr /1972

(a)   ex Grimethorpe Colliery, /1960
(b)   ex Hemsworth Colliery, c2/1955

(1)   written off by 7/1978

Gauge : 2ft 0in (or 2ft 1in?) (Surface stockyard)

-	4wDM	HE	7274	1973	New	(1)

(1)   to Houghton Main Colliery, 30/8/1988

Gauge : 200mm (Becorit Roadrailer trapped rail system)

No.2		1adDHF	BGB	DRL25/2/213	1970		
		frame built HE				New	(1)
No.1		1adDHF	BGB	DRL25/ /	1971	New	(1)
No.3	70/117	2adDHF	BGB	DRL40/1/506	1972	New	(3)
No.4		2adDHF	BGB	DRL40/3/513	1973	New	(2)
No.5		2adDHF	BGB	DRL40/3/517	1973	New	(2)
No.6	70/426	2adDHF	BGB	DRL50/200/527	1975	New	(3)
No.7	70/19992	2adDHF	BGB	DRL50/200/536	1976	New	(3)
No.11		2adDHF	BGB	DRL50/200/523	1975	(a)	(3)
No.12		2adDHF	BGB	DRL50/200/531	1976	(a)	(3)

(a)   ex Swadlincote Machinery Stores, Derbyshire, c2/1985

(1)   written off, 4/1974 and scrapped, c/1974
(2)   written off, 1/1977 and scrapped, c/1977
(3)   sold, scrapped or abandoned underground, by 1993

DARTON COLLIERY, Darton   N23
ex **Fountain & Son Ltd**   SE 323097
NE6 from 1/1/1947   CLOSED 8/1948
See NORTH GAWBER COLLIERY

DARTON DISPOSAL POINT, Darton   N24
                                                                         SE 306103
Opened by MFP c1944; OE from 1/4/1952.   CLOSED 1960

The Disposal Point was located on the site of Darton Main Colliery (which had closed in the 1930s).
Sidings ran west from the BR (ex LMSR) line, north of Darton Station, for ¼ mile.
Operated by contractors :   Initially by **Sir Lindsay Parkinson & Co Ltd**
                                 Latterly **by William Pepper & Co Ltd.**

Gauge : 4ft 8½in

SLP 44	HANDY	0-6-0ST	IC	BH	1115	1896	(a)	(1)
71499		0-6-0ST	IC	HC	1776	1944	New (b)	(2)
	REGINA	0-6-0ST	IC	HC	466	1897	(c)	(3)
(92)	PEPPER	0-4-0T	OC	9E		1892	(d)	(4)
	BRAMLEY No.4	0-6-0ST	OC	HE	1643	1929		
		reb		HE		1938	(e)	(5)
206	ALLENBY	0-6-0ST	IC	MW	1379	1898	(f)	(6)
	HOYLAND	0-4-0ST	OC	YE	1026	1910	(g)	(7)
No.1	QUEEN	0-4-0ST	OC	YE	1027	1910	(h)	(8)

(a)   reported as here, c/1946;
        believed to Fountain & Burnley Ltd, Haigh Colliery, loan, (c12/1946 ?); returned
(b)   loan from War Department;
        to Garswood Hall Collieries Co Ltd, Lancashire, loan, c/1945 (by 8/1945); returned
(c)   ex Denby Grange Collieries Ltd (possibly on loan), c/1946
(d)   ex BR, 5/1949;
        to British Oak Disposal Point, West Yorkshire, 10/11/1949;
        ex Skiers Spring Disposal Point, after 3/1950, by 5/1952
(e)   ex HE, Leeds, West Yorkshire (hire), c/1949 (by 3/1950)
(f)   ex Sir Lindsay Parkinson & Co Ltd, Temple Newsam Plant Depot, Leeds,
        after 24/2/1948, by 17/7/1949;
        (possibly then to other sites and returned here before disposal)
(g)   ex Wath - Elsecar Disposal Point /1957;
        to British Oak Disposal Point, West Yorkshire, 1/1958; returned, 4/1958
(h)   ex British Oak Disposal Point, West Yorkshire, 1/1958

(1)   to Sir Lindsay Parkinson & Co Ltd, Temple Newsam Plant Depot, Leeds, by 24/2/1948
(2)   to Peel Hall Disposal Point, Lancashire, c/1948 (by 3/1948)
(3)   (either returned to Denby Grange Collieries Ltd, 7/1947 or remained here until);
        to HC, Leeds, West Yorkshire, 5/1948 and scrapped there, 6/1948
(4)   to Wm. George (Wath) Ltd, Wath-on-Dearne, 4/1961 and scrapped there, 6/1961
(5)   returned to HE, off hire, after 3/1950, by 13/6/1950
(6)   to Sir Lindsay Parkinson & Co Ltd, after 3/1950, by 3/1951
(7)   scrapped on site by Thos.W.Ward Ltd, /1961 (after 9/1961)
(8)   to Wath - Elsecar Disposal Point , /1958 (after 4/1958)

DEARNE VALLEY COLLIERY, Little Houghton   N25
ex Dearne Valley Colliery Co Ltd   SE 423054

NE4 from 1/1/1947: NE5 from 26/4/1964; BNY from 26/3/1967; NYK from 1/10/1985; NYG from 1/4/1990.   CLOSED 8/3/1991

The colliery was served by sidings on the east side of the BR (ex LMSR) line, ¾ miles north of Darfield Station. Shunting was by main line locos except during an industrial dispute in 1955 and for an unspecified reason in 1967-1969. A surface conveyor was installed (possibly in the late 1970s) alongside the BR line to convey Dearne Valley coal to Houghton Main Washery. Rail traffic ceased by 1988.

Gauge: 4ft 8½in

VICTORY	0-6-0ST	IC	P	1519	1919	(a)	(1)
-	0-4-0DM		JF	22558	1939	(b)	(2)

(a)   ex Houghton Main Colliery, 5/1955
(b)   ex Barnsley Main Colliery, after 7/1967, by 4/1968

(1)   to Houghton Main Colliery, 7/1955
(2)   to Park Mill Colliery, West Yorkshire, 25/8/1969

DENABY COLLIERY, Mexborough
ex **Amalgamated Denaby Collieries Ltd**

NE3 from 1/1/1947; SYK from 26/3/1967.   Merged with CADEBY and surface CLOSED 3/1968

The colliery was located on the north side of the BR (ex LNER) Mexborough - Doncaster line (1½ miles west of Conisborough Station) to which it was connected at Denaby Crossing and Lowfield Junction. There was also a spur from the former LMSR Dearne Valley Railway which passed north of the colliery and adjacent to which extensive dirt tips were established, served both by rail and aerial ropeway. Close to Lowfield Junction, there was a connection with the Wrangbrook Junction line also used by NCB traffic to and from Cadeby Colliery. A staith on the River Don, east of Denaby, was also used by Cadeby Colliery until 1981. Narrow gauge locomotives were used in a stock yard on the north side of the river which was linked to the pit top by a gantry. Coal winding at Denaby ceased in 1959 but the loco shed continued to be used for repairs and to store surplus Cadeby locomotives. The surface, except for the staith and the wagon works which were used and shunted by Cadeby, closed in 1968. The remainder followed in 1981. The underground locos were used on coal haulage until 1968

Gauge : 4ft 8½in

A combined list of standard gauge locomotives used at either Cadeby or Denaby Collieries will be found under the Cadeby heading, notwithstanding that in the 1950s most locomotives were at Denaby.

Gauge : 1ft 11in (Surface stockyard)

This system was an approximately circular formation, which crossed the river on two separate bridges.

-	4wDM	MR	7606	1939	(a)	(1)
-	4wDM	MR	8814	1943	(b)	(2)

(a)   ex George W. Bungey Ltd, dealer, Hayes, Middlesex, after 9/1950, by 9/1952;
       earlier at Durham County Council
(b)   ex unknown location, early 1950s;
       originally WD and then with Wm Jones Ltd, dealer, Greenwich, London, 23/6/1947.

(1)   to Cadeby Colliery, via underground connection, c/1966 (by 11/1966)
(2)   to Cadeby Colliery, /1965

Gauge : 1ft 11in (Underground locomotives)

No.1	0-4-0DMF	HE	4129	1950	New	(1)
No.2	0-4-0DMF	HE	4130	1950	New	(1)
No.3	0-4-0DMF	HE	4332	1950	New	(1)
No.4	0-4-0DMF	HE	4333	1950	New	(1)
No.5	0-4-0DMF	HE	4808	1954	New	(1)
No.6	0-4-0DMF	HE	5203	1957	New	(1)
-	4wDMF	RH	268860	1949	(a)	(1)

(a)   ex Elsecar Central Workshops, 8/1961

(1)   to Cadeby Colliery, via underground connection, /1968 (by 11/1968)

DINNINGTON COKING PLANT, Dinnington
ex **Amalgamated Denaby Collieries Ltd**

NE(C) from 1/1/47; CPD from 1/1/1963.
Adjacent to DINNINGTON COLLIERY - which see.

N28

ex **Amalgamated Denaby Collieries Ltd**   SK 517867

NE1 from 1/1/1947; SYK from 26/3/1967; SYG from 1/4/1990.   CLOSED 10/1991

Located at the end of a BR branch which ran north-east for ¾ mile from a junction ¾ mile south of Dinnington & Laughton Station. The adjacent DINNINGTON COKING PLANT was shunted by the colliery locomotive. Use of standard gauge locomotives ceased in 1986. The underground locomotives were used intermittently for supplies.

Gauge: 4ft 8½in

	DINNINGTON No.1	0-4-0ST	OC	LE	242	1907	(a)	(1)
	DINNINGTON No.2	0-4-0ST	OC	HC	862	1909		
	latterly carried plates HC				989		(a)	(2)
	ROSSINGTON No.1	0-4-0ST	OC	HC	989	1912	(b)	(3)
	MALTBY No.2	0-4-0ST	OC	HL	3910	1937	(c)	(4)
	DINNINGTON No.3	4wVBT	VCG	S	9526	1951	New (d)	(5)
	MALTBY No.1	0-4-0ST	OC	HL	3771	1930	(e)	(6)
No.23	(CLARRIE No.54)	0-6-0DM		HC	D1090	1958	(f)	(7)
No.11	521/11							
	DINNINGTON (No.1)	0-6-0DM		HC	D1113	1958	New	(8)
	BILL	0-6-0ST	OC	AE	1920	1924	(g)	(9)
	D2607	0-6-0DM		HE	5656	1960	(h)	(10)
No.31	(D2328)	0-6-0DM		RSHD	8187	1961		
	(DINNINGTON No.2)			DC	2709	1961	(j)	(11)
No.12	D2327	0-6-0DM		RSHD	8186	1961		
	(No.2) 521/12			DC	2708	1961	(k)	(12)
No.33	(D2334)	0-6-0DM		RSHD	8193	1961		
				DC	2715	1961	(m)	(13)
	D2332 LLOYD	0-6-0DM		RSHD	8191	1961		
				DC	2713	1961	(n)	(14)

(a)   ex Amalgamated Denaby Collieries Ltd, with site, 1/1/1947
(b)   ex Amalgamated Denaby Collieries Ltd, with site, 1/1/1947;
        to Waleswood Coking Plant, after 8/1959, by 8/1960; returned, after 10/1961, by 10/1962
(c)   ex Maltby Colliery, after 4/1950, by 24/9/1950
(d)   to Shireoaks Colliery, Nottinghamshire, after 8/1951, by 29/8/1952;
        returned after 4/1953, by 10/1953
(e)   ex Nunnery Colliery, 5/9/1956
(f)   ex New Stubbin Colliery, 10/2/1978; after 2/1978, by 5/1979
        to Manton Colliery, Nottinghamshire, 15/2/1978; returned, after 2/1978, by 5/1979
(g)   ex Firbeck Colliery, Nottinghamshire, after 8/1959, by 9/1960
(h)   ex BR, Llandudno Junction, Caernarvonshire, 28/9/1968
(j)   ex BR, Gateshead, 6/6/1969
(k)   ex Manton Colliery, Nottinghamshire, 9/8/1971;
        to Elsecar Central Workshops, 3/5/1973; returned;
        to Elsecar Central Workshops, 15/11/1974; returned, 20/1/1975
(m) ex Thurcroft Colliery, after 21/5/1985, by 19/6/1985
(n)   ex Thurcroft Colliery, after 19/7/1985, by 9/1985

(1)   scrapped on site by W. Robinson & Co (Sheffield) Ltd, of Sheffield, 4/1955
(2)   to Steetley Colliery, Nottinghamshire, after 6/1965, by 3/1966
(3)   dismantled by 8/6/1963; scrapped by 3/1964
(4)   to Maltby Colliery, after 24/9/1950, by 26/5/1951
(5)   to Manton Colliery, Nottinghamshire, 3/1959
(6)   to Maltby Colliery, after 9/1956, by 4/1958
(7)   to Booth Roe Metals Ltd, Rotherham, 5/2/1988 and scrapped there, 2/1988

(8) scrapped on site (by an unknown contractor), first week of 10/1986
(9) written off by 16/5/1970; scrapped on site after 11/1970, by 4/1972
(10) to Steetley Colliery Nottinghamshire, after 9/1968, by 11/1968
(11) to Steetley Colliery, Nottinghamshire, 4/1973
(12) to Coopers (Metals) Ltd, Sheffield, 5/1/1984, and scrapped 2/1984
(13) to Maltby Colliery, 24/2/1986
(14) scrapped on site (by an unknown contractor), first week of 7/1986

Gauge : 2ft 0in (Surface stockyard)

No.1	524/11	4wBEF	CE	B1574E	1978	New	(1)
No.2	524/12	4wBEF	CE	B1574F	1978	New	(2)
No.22	No.21 390/16201	0-4-0DMF	HE	4114	1955	(a)	(3)
No.28		0-4-0DMF	HE	4503	1955	(a)	(4)
No.52		0-4-0DMF	HE	5514	1958	(a)	(4)

(a) ex New Stubbin Colliery, c4/1979

(1) to underground use, /1986 (after 29/1/1986).
(2) to CE, Hatton, Derbyshire, after 27/2/1992, by 20/5/1992;
     thence to Houghton Main Colliery
(3) possibly never used here; out of use, 3/1980; scrapped c/1984 (after 16/4/1983)
(4) possibly never used here; out of use, 3/1980; sold for scrap, 3/3/1981

Gauge : 2ft 0in (Underground locomotives)

-		0-4-0DMF	HE	3344	1946	(a)	(1)
	-	0-4-0DMF	HC	DM741	1951	New	(2)
No.3	524/13	4wBEF	CE	B1574C	1978	New	(3)
	-	4wBEF	CE	B3038	1983	New (b)	(4)
No.1	524/11	4wBEF	CE	B1574E	1978	(c)	(5)

(a) ex Amalgamated Denaby Collieries Ltd, with site, 1/1/1947
(b) to CE, Hatton, Derbyshire, /1989; returned 10/1989
(c) ex surface use, after 6/1986

(1) believed abandoned underground, c/1969
(2) to Kiveton Park Colliery 7/1964
(3) to Thurcroft Colliery 7/1981
(4) to CE, Hatton, Derbyshire, 11/1991; altered to 2ft 3½in gauge and thence to Bentley Colliery
(5) to CE, Hatton, Derbyshire, 11/1991; thence to Manton Colliery, Nottinghamshire

DODWORTH COLLIERY, Dodworth   N29
ex Old Silkstone Collieries Ltd   SE 312058

NE5 from 1/1/1947; BNY from 26/3/1967; NYK from 1/10/1985.
Merged with REDBROOK COLLIERY and surface CLOSED 1987

The colliery was served by sidings which ran north from the BR (ex LNER) line, north-east of Dodworth Station. Use of surface locomotives had ceased by 1986. A second mine, HIGHAM COLLIERY (SE 309069) was located one mile to the north and was linked to Dodworth by overland conveyor. It was part of Dodworth and had no rail connection in the NCB period.

Gauge : 4ft 8½in

No.1	CECIL LEVITA	0-6-0ST	IC	HE	1499	1926	(a)	Scr 7/1961
	BRAMLEY No.4	0-6-0ST	IC	HE	1643	1929		
		rep	HE		1938	(b)	(1)	

(75008)		0-6-0ST	IC	HE	2857	1943		
			rep	HE	59295	1966	(c)	(2)
	AVON No.3	0-6-0ST	OC	AE	1826	1919	(d)	(3)
	H.C. No.2	0-4-0ST	OC	HC	1890	1960	New (e)	(4)
(No.1)		0-6-0ST	IC	P	1518	1919	(f)	(5)
	H.C. No.1	0-4-0ST	OC	HC	1889	1960	(g)	(4)
TL3		4wDH		TH	158C	1965		
	built on frame of			S	9557		(h)	(6)
(2239)		0-6-0DM		VF	D289	1956		
				DC	2563	1956	(j)	(7)
-		4wDH		S	10058	1961	(k)	(8)

(a)   ex Old Silkstone Collieries Ltd, with site, 1/1/1947
(b)   ex HE, Leeds, West Yorkshire, hire, 8/1947
(c)   ex War Department (possibly on loan to MFP), /1948 (by 7/1948);
       to HE, Leeds, West Yorkshire, 30/4/1965; returned after 7/1966, by 2/1967
(d)   ex Wombwell Colliery, after 4/1957, by 7/1958;
       to Wombwell Colliery, after 2/1959, by 10/1959; returned 10/1959 or 11/1959
(e)   to HE, Leeds, West Yorkshire, /1966; returned, /1966
(f)   ex Barrow Colliery, after 11/1964, by 4/1965
(g)   ex Monk Bretton Colliery 2/10/1968
(h)   ex Wombwell Colliery, after 6/1969, by 2/1970
(j)   ex BR Selhurst, London, 9/1972
(k)   ex TH, Kilnhurst (by rail under own power), 18/4/1974

(1)   to John Baker & Bessemer Ltd, Kilnhurst (further hire from HE), 9/1947
(2)   to Cadley Hill Colliery, Derbyshire, 1/1976 (in transit, 23/1/1976)
(3)   to Barrow Colliery, after 11/1959, by 12/3/1961
(4)   scrapped on site during week ending 8/3/1975
(5)   scrapped on site by Thos.W.Ward Ltd, 7/1967
(6)   to Barrow Colliery, after 8/1970, by 3/1971
(7)   to C.F. Booth Ltd, Rotherham, 20/3/1986; and scrapped there, by 27/3/1986
(8)   to C.F. Booth Ltd, Rotherham, 19/3/1986; and scrapped there, 3/1987 (after 12/3/1987)

Gauge : 1ft 9in (Surface system)

-	4wDM	HE	6631	1965	New	(1)

(1)   to Park Mill Colliery, West Yorkshire (for repairs), 10/1985

ELSECAR CENTRAL WORKSHOPS, Elsecar   N30
ex Earl Fitzwilliam's Collieries Co
SK 387999

NE3 from 1/1/1947; SYK from 26/3/1967; HQ from 6/1967.   CLOSED after 1975

Workshops located adjacent to Elsecar Goods Station at the terminus of the BR (ex LNER) line from Elsecar Junction.

Gauge: 4ft 8½in

No.8	SUCCESS	0-6-0ST	OC	FW	382	1878	(a)	(1)
No.3		0-6-0ST	OC	HC	285	1889	(b)	(2)
No.29	(RICHARD)	0-4-0ST	OC	MW	1968	1919	(c)	(3)
No.11		0-6-0ST	OC	HC	1368	1920	(d)	(4)
(No.2)		0-6-0ST	OC	HC	1052	1914	(e)	(5)
No.8	D.M.C.No.2	0-4-0ST	OC	P	701	1898	(f)	(6)
(No.19)	(FITZWILLIAM No.7)	0-4-0ST	OC	HC	916	1910	(g)	(7)
No.40	TINSLEY	0-6-0ST	OC	AB	2025	1936	(h)	(8)

	ATLAS No.6	0-4-0ST	OC	YE	478	1892	(j)	(9)
No.27		0-4-0ST	OC	HC	1338	1918	(k)	(10)
	MANVERS MAIN 42	0-6-0T	IC	HC	1690	1937	(m)	(11)
44		0-6-0ST	OC	P	1891	1940	(n)	(12)
No.17	(Mo.3)	0-4-0ST	OC	HC	751	1906	(p)	(13)
No.30		0-6-0ST	OC	AE	1819	1919	(q)	(14)
No.31		0-6-0T	OC	AB	1717	1921	(r)	(15)
No.10	ATLAS No.15	0-4-0ST	OC	HL	2464	1900	(s)	(16)
	T.C.D.	0-4-0ST	OC	YE	1891	1923	(t)	(17)
No.9		0-4-0ST	OC	HL	2454	1900	(u)	(18)
No.38		0-6-0ST	IC	HL	3658	1926	(v)	(19)
No.35		0-6-0ST	OC	AB	1792	1923		
		reb		AB		1938	(w)	(20)
	DAVID No.58	0-6-0DM		HC	D1128	1958	(x)	(21)
No.53		0-6-0ST	IC	HE	3856	1956	(y)	(22)
No.43	JENKY	0-6-0ST	IC	RSHN	6942	1938	(z)	(23)
No.63	(68067)	0-6-0ST	IC	HC	1792	1946	(aa)	(24)
	TERRY No.56	0-6-0DM		HC	D1091	1958	(ab)	(25)
No.23		0-6-0ST	IC	HC	1753	1943	(ac)	(26)
No.12	521/12 (D2327)	0-6-0DM		RSHD	8186	1961		
				DC	2708	1961	(ad)	(27)

(a) ex Elsecar Colliery, 7/1947
(b) ex Manvers Colliery, 11/1947
(c) ex Furness Shipbuilding & Engineering Co Ltd, Haverton Hill, Cleveland, 9/1948,
     per H. Potter & Co Ltd, dealers (Harold Potter & Co Ltd of Nottingham ?))
(d) ex Appleby-Frodingham Steel Co Ltd, Scunthorpe, Lincolnshire, 1/1950
(e) ex Appleby-Frodingham Steel Co Ltd, Scunthorpe, Lincolnshire, 3/1950
     (note that it may have been delivered to Elsecar Colliery rather than the Central Workshops);
     to Wath Colliery, 6/1952; returned, /1953
(f) ex Silverwood Colliery, 2/1953
(g) ex Rotherham Main Colliery, 9/1952;
     to Elsecar Colliery, 4/1953; returned, after 7/1956, by 4/1957;
     to Aldwarke Main Colliery, 6/2/1958; ex Elsecar Colliery, by 12/3/1961;
     to Elsecar Colliery, after 3/1961, by 17/6/1961; returned after 7/1961, by 23/2/1963;
     to Elsecar Colliery, after 2/1963, by 5/5/1963; returned, by 5/1964
(h) ex Kilnhurst Colliery after 8/1953, by 9/1954
(j) ex Rotherham Main Coking Plant, c10/1954 (after 8/1954)
(k) ex Elsecar Colliery, after 8/1953, by 7/1955;
     to Cortonwood Colliery, /1955; ex Elsecar Colliery, after 12/3/1961, by 13/5/1961
(m) ex Manvers Colliery, /1956 (by 7/1956)
(n) ex Manvers Colliery, after 7/1956 by 4/1957
(p) ex Silverwood Colliery, by 3/1952;
     to Roundwood Staith, 12/1952; ex Aldwarke Main Colliery 7/1957
(q) ex Cortonwood Colliery, 3/1956
(r) ex Silverwood Colliery, 20/8/1958
(s) ex Aldwarke Main Colliery, after 4/1958, by 2/1959
(t) ex Smithywood Coking Plant, c5/1959 (by 5/1959)
(u) ex Rotherham Main Coking Plant c1/1960 (possibly 29/2/1960)
(v) ex Cadeby Colliery, after 5/1959, by 12/3/1961
(w) ex Cortonwood Colliery, after 4/1960, by 12/3/1961
(x) ex Elsecar Colliery, /1963
(y) ex Cortonwood Colliery, 15/1/1963
(z) ex Denaby Colliery, after 10/1962, by 23/2/1963
(aa) ex Manvers Colliery, after 11/1963, by 3/1964

(ab)   ex Silverwood Colliery, after 4/1965, by 7/1965
(ac)   ex Silverwood Colliery, after 4/1966, by 9/1966
(ad)   ex Dinnington Colliery, 3/5/1973;
         to Dinnington Colliery, /1973; returned, 15/11/1974

(1)   to New Stubbin Colliery, 9/1948
(2)   to Wath Colliery, 9/1948
(3)   to Aldwarke Main Colliery, 1/1949
(4)   to New Stubbin Colliery, after 27/5/1950, by 9/1950
(5)   dismantled for spares and remainder scrapped, 6/1956
(6)   to Silverwood Colliery, by 8/1953
(7)   dismantled 5/1964; scrapped by 10/1964
(8)   to Kilnhurst Colliery, after 9/1954, by 11/1954
(9)   to Rotherham Main Coking Plant, c2/1955 (by 5/1955)
(10)   to Elsecar Colliery, /1961 (after 13/5/1961)
(11)   to Manvers Colliery, c/1958
(12)   to Manvers Colliery, after 4/1957, by 4/1958
(13)   to Aldwarke Main Colliery, after 7/1957, by 9/1958
(14)   scrapped, after 7/1956
(15)   to Elsecar Colliery 9/1959
(16)   to Aldwarke Main Colliery, 26/2/1960
(17)   to Smithywood Coking Plant c3/1960 (by 9/4/1960)
(18)   to Rotherham Main Coking Plant, c4/1960
(19)   to Denaby Colliery, after 10/1961, by 7/1962
(20)   to Cortonwood Colliery, after 16/7/1961, by 12/1962
(21)   to Elsecar Colliery, c/1963
(22)   to Cortonwood Colliery, after 6/1965, by 10/1965
(23)   scrapped, c/1963 (after 5/1963)
(24)   to Manvers Colliery, after 10/1965 by 4/1966
(25)   to Silverwood Colliery, /1966 (after 4/1966)
(26)   scrapped, c10/1968 (after 9/1968)
(27)   to Dinnington Colliery, 20/1/1975

Gauge : 3ft 0in

No.44		0-6-0DMF	HE	5314	1956	(a)	(1)

(a)   ex Silverwood Colliery, c/1962

(1)   to Silverwood Colliery, 8/10/1962

Gauge : 2ft 0in

-		4wDM	RH	192843	1938	(a)	(1)
No.5		4wDM	RH	256275	1948	(b)	(2)
-		4wDM	RH	249567	1947	(b)	(2)
-		4wDM	MR	7218	1938	(c)	(3)
No.13	390/12101	0-4-0DMF	HE	3557	1948	(d)	(4)

(a)   ex Wentworth Drift Mine, date unknown
(b)   ex Wentworth Drift Mine, c3/1956

(c) ex unknown location;
originally Fletcher & Co (Contractors) Ltd, (of Mansfield, Nottinghamshire), Edgware contract, Middlesex.
(d) ex Elsecar Colliery, 10/1958

(1) sold or scrapped after 2/3/1949
(2) to Aldwarke Colliery, (1ft 11in gauge), after 5/1956, by 1/6/1957
(3) to New Stubbin Colliery, /1957
(4) to Elsecar Colliery c/1959

Gauge : 1ft 11in

| | | 4wDM | RH | 268860 | 1949 | (a) | (1) |

(a) ex Aldwarke Colliery, 5/1961

(1) to Denaby Colliery 8/1961

ELSECAR COLLIERY, Elsecar   N31
ex Earl Fitzwilliam's Collieries Co
SE 390003

NE3 from 1/1/1947; SYK from 26/3/1967; Rail traffic ceased 1981.   CLOSED 10/1983

Served by sidings on the east side of the BR (ex LNER) Elsecar Goods branch, ½ mile north of Elsecar Goods Station. A line also ran north from the colliery to a dirt tip (½ mile). Rail traffic ceased and connection lifted by 11/1981. Locomotives were used underground.

Gauge : 4ft 8½in

No.	Name	Type		Mkr	No.	Date		
No.3	(MILTON No.2)	0-4-0ST	OC	YE	119	1869	(a)	(1)
No.19	FITZWILLIAM (No.7)	0-4-0ST	OC	HC	916	1910	(b)	(2)
No.27	(No.10)	0-4-0ST	OC	HC	1338	1918	(c)	(3)
No.8	SUCCESS	0-6-0ST	OC	FW	382	1878	(d)	(4)
No.3	(WENTWORTH)	0-4-0ST	OC	YE	120	1869	(e)	(5)
No.31		0-6-0T	OC	AB	1717	1921	(f)	(6)
No.22	DAVID No.58	0-6-0DM		HC	D1128	1958	New (g)	(7)
	ATLAS No.6	0-4-0ST	OC	YE	478	1892	(h)	(8)
No.13	WALTER No.69							
	521/11001	0-6-0DH		HE	6230	1963	(j)	(9)
No.14	WILF No.72							
	521/11944	0-6-0DH		HE	6287	1965	New	(10)
No.34		0-6-0T	OC	HC	1523	1925	(k)	(11)

(a) ex Earl Fitzwilliam's Collieries Co, with site,1/1/1947
(b) ex Earl Fitzwilliam's Collieries Co, with site,1/1/1947;
to Rotherham Main Colliery, 10/1949; ex Elsecar Central Workshops, 4/1953;
to Elsecar Central Workshops, after 7/1956 by 4/1957; ex Aldwarke Main Colliery, 6/8/1958;
to Elsecar Central Workshops, after 7/1961, by 2/1963; returned, by 5/5/1963
(c) ex Earl Fitzwilliam's Collieries Co, with site,1/1/1947;
to Elsecar Central Workshops, after 8/1953 by 7/1955; ex Cortonwood Colliery, /1955;
to Manvers Colliery, 29/6/1956; returned, 8/1956;
to Elsecar Central Workshops, after 12/3/1961, by 13/5/1961; returned, /1961 (after 13/5/1961)
(d) ex New Stubbin Colliery, 3/1947
(e) ex New Stubbin Colliery, 2/1950
(f) ex Elsecar Central Workshops, 9/1959
(g) to Elsecar Central Workshops, /1963; returned, c/1963
(h) ex New Stubbin Colliery, c1/1964 (after 7/1963, by 3/1964)

(j)   ex HE, Leeds, West Yorkshire, /1963: earlier a demonstration locomotive at locations
     including Waterloo Main Colliery, West Yorkshire
(k)   ex New Stubbin Colliery, after 4/1968, by 9/1968

(1)   to Aldwarke Main Colliery, after 3/1959, by 8/1959
(2)   to Elsecar Central Workshops, by 5/1964
(3)   scrapped after 9/1966, by 6/1967
(4)   to Elsecar Central Workshops, 7/1947
(5)   to New Stubbin Colliery, after 5/1950, by 8/1950
(6)   to Silverwood Colliery, 9/1959
(7)   to New Stubbin Colliery, after 9/1966, by 8/1967
(8)   scrapped on site by Frank Green & Sons Ltd, of Stairfoot, week ending 13/3/1966
(9)   to Silverwood Colliery, 18/3/1981
(10) to Cortonwood Colliery, 9/1981
(11) scrapped, c3/1970 (by 16/5/1970)

Gauge : 2ft 0in

No.13	390/12101	0-4-0DMF	HE	3557	1948	New (a)	(1)
No.14	390/12102	0-4-0DMF	HE	3558	1948	New	(2)
No.23	390/12100	0-4-0DMF	HE	4128	1949	New	(3)
No.34		0-4-0DMF	HE	4025	1949	(b)	(3)
No.60	390/12099	0-4-0DMF	HE	5423	1964	New	(4)
No.56	390/12778	0-4-0DMF	HE	5597	1961	New	(5)
No.57	390/12779	0-4-0DMF	HE	5598	1961	New	(4)
No.55		0-4-0DMF	HE	6059	1962	(c)	(6)

(a)   to Elsecar Central Workshops, 10/1958; returned, c/1959
(b)   ex Maltby Colliery, 11/1957
(c)   ex New Stubbin Colliery, for spares, 7/1979

(1)   to Shireoaks Colliery, Nottinghamshire, c/1970
(2)   to New Stubbin Colliery, 7/1975
(3)   written off, after 10/1983
(4)   to Kiveton Park Colliery, 1/1984
(5)   to Shireoaks Colliery, Nottinghamshire, 5/8/1975
(6)   remains to Barnburgh Colliery, 26/3/1980

FENCE CENTRAL WORKSHOPS, Woodhouse   N32
SK 437857

Opened by NE1 c/1960; SYK; from 26/3/1967; HQ from 6/1967; National Plant Centre from 1989.
CLOSED by 1994

Located on the site of the closed Fence Colliery. Served by sidings on the east side of the BR (ex
LMSR) line south of Woodhouse Mill Station.

Gauge : 4ft 8½in

D2607	521/21	0-6-0DM	HE	5656	1960	(a)	(1)

(a)   ex Steetley Colliery, Nottinghamshire, 28/5/1974

(1)   ex Steetley Colliery, Nottinghamshire, 30/10/1974

Gauge : 2ft 2in

No.73	390/7223	0-4-0DMF	HE	#	1957		
		reb	HE	7223	1971	(a)	(1)

Either HE 4332 or HE 5203 – see note in Cadeby Colliery entry.

(a)   ex Wath Colliery, /1971 (by 7/1971)

(1)   to Barnburgh Main Colliery 22/7/1971

FERRYMOOR COLLIERY, Grimethorpe N33
ex **Hodroyd Coal Co Ltd**   SE 408088

NE4 from 1/1/1947; NE6 from 26/4/1964; BNY from 26/3/1967.
Absorbed RIDDINGS DRIFT (West Yorkshire) 4/1973; NYK from 1/10/1985; Merged with SOUTH
KIRKBY 3/1987.   Complex CLOSED 3/1988.

Located on the west side of GRIMETHORPE COLLIERY with which surface installations and sidings
were merged in 1947 and which see for further details. At vesting day, a narrow gauge (possibly rope-
worked) tubway ran north from Ferrymoor to BRIERLEY COLLIERY but the latter closed during
1/1947.

Note that a number of flameproof Clayton and Hunslet locomotives delivered to
FERRYMOOR/RIDDINGS COLLIERY during 1879 – 1985 were all for the Riddings section and thus
have been listed in the NCB West and North Yorkshire volume already published.

Gauge : 4ft 8½in

674	HODROYD		0-6-0ST	IC	RS	2893	1898	(a)	(1)

(a)   ex Hodroyd Coal Co Ltd, with site, 1/1/1947

(1)   to Grimethorpe Colliery, /1947

GOLDTHORPE COLLIERY, Goldthorpe N34
ex **Goldthorpe Collieries Ltd**   SE 469042

NE2 from 1/1/47; DCR from 26/3/1967; NYK from 1/10/1985;
NYG from 1/4/1990; SYG from 1/10/1991.   CLOSED 2/1994

Served by sidings on the north east side of the BR (ex LMSR) Dearne Valley Railway, 1¼ miles north-
west of Harlington Halt. Shunted by main line locomotives. Much of the Dearne Valley line closed in
1966, including the section south of Goldthorpe. Thereafter, changes in 1978 saw Goldthorpe
accessed via a new connection to Hickleton Colliery (which see). It is not known if the underground
locomotive was put to use.

Gauge : 2ft 0in (Underground locomotive)

	0-6-0DMF	HC	DM1151	1959	New	(1)

(1)   to Rossington Colliery, 28/8/1960 (possibly without being used here)

GRANGE COLLIERY, Kimberworth N35
ex **N.C. Thorncliffe Collieries Ltd**   SK 395940

NE3 from 1/1/1947.   CLOSED 7/1962

Colliery sidings were at the terminus of the BR (ex LNER) Dropping Well branch which ran south east
and then north east for 2 miles from Grange Lane Station. Use of a loco ceased in 1960.

Gauge : 4ft 8½in

("OWD BUG")	4wPM	NC		c1939	(a)	(1)
VENTURE	0-4-0DM	JF	22287	1938	(b)	(2)
-	0-4-0DM	JF	22558	1939	(c)	(3)

(a)   ex N.C. Thorncliffe Collieries Ltd, with site, 1/1/1947
(b)   ex Thos.W.Ward Ltd, Sheffield c/1954;
       earlier Bold Venture Lime Co Ltd, Chatburn, near Clitheroe, Lancashire;
       to Barrow Colliery, after /1954, by 4/1955; ex Birdwell Central Workshops, after 29/4/1956

(c)   ex JF, Leeds, West Yorkshire, 5/1951;
earlier NCB, Caerphilly Tar Works, Glamorgan

(1)   scrapped c/1950
(2)   to Birdwell Central Workshops, by 3/1957
(3)   to Smithy Wood Colliery, after 10/1955, by 9/4/1960

GRIMETHORPE COLLIERY, Grimethorpe   N36

ex **Carlton Main Coal Co Ltd**   Loco shed - SE410085 (later SE 406083)

NE4 from 1/1/1947; NE6 from 26/4/1964; BNY from 26/3/1967;
NYK from 1/10/1985; NYG from 1/1/1990; SYG from 1/10/1991.   CLOSED 7/5/1993.

Sidings ran north from the end of a BR branch which left the former LMSR line 1 mile south-east of Cudworth Station and ran north-east to the colliery. It was also connected to the Dearne Valley line where it crossed the sidings and to the terminus of a BR (ex LNER) branch which ran east from Stairfoot. In the 1950s a new connection was made by BR to replace the earlier lines. This ran north and then north-west from the BR (ex LMSR) line, 1½ miles south-east of Cudworth Station, and also served Houghton Main Colliery. FERRYMOOR COLLIERY was located on the Grimethorpe site and the surfaces merged in 1947 No standard gauge locomotives from 1986. The underground locomotives were used to haul coal to the shaft until replaced by conveyors c1978.

Gauge : 4ft 8½in

No.2		0-6-0ST	IC	BP	4392	1901	(a)	(1)
No.1		0-6-0T	IC	K	5182	1919	(a)	(2)
No.3		0-6-0ST	IC	HE	2704	1945	(a)	(3)
No.4		0-6-0ST	IC	VF	5295	1945		
			rep	HE	59289	1966	(b)	(4)
No.5		0-6-0ST	IC	VF	5296	1945		
			rep	HE	59296	1967	(c)	(5)
674	HODROYD	0-6-0ST	IC	RS	2893	1898	(d)	(6)
No.2	(75128)	0-6-0ST	IC	HE	3178	1944	(e)	(7)
	YORK No.2	0-4-0ST	OC	YE	2473	1949	(f)	(8)
	-	0-4-0DH		HC	D1259	1962	(g)	(9)
	-	0-4-0DM		HC	D1094	1959	(h)	(10)
No.1	(D2057)	0-6-0DM		Don		1959		
		reb to 0-6-0DMF		HE	6643	1967	(j)	(11)
No.2	(D2093)	0-6-0DM		Don		1959		
		reb to 0-6-0DMF		HE	6645	1967	(j)	(12)
	D2284	0-6-0DM		RSHD	8102	1960		
				DC	2661	1960	(k)	(13)

(a)   ex Carlton Main Colliery Co Ltd, with site, 1/1/1947
(b)   ex Carlton Main Colliery Co Ltd, with site, 1/1/1947;
       to HE, Leeds, West Yorkshire, 4/1965; returned, /1966
(c)   ex Carlton Main Colliery Co Ltd, with site, 1/1/1947;
       to Frickley Colliery, West Yorkshire, after 4/1949, by 9/1949;
       returned 10/1949 (after 12/10/1949)
       to HE, Leeds, West Yorkshire, 4/1965, by 6/1965; returned, 2/1967
(d)   ex Ferrymoor Colliery, /1947
(e)   ex Houghton Main Colliery, after 9/9/1951, by 4/1953
(f)   ex Barrow Colliery, 1966 (after 4/1966)
(g)   ex Crigglestone Colliery, West Yorkshire, 23/8/1968
(h)   ex Shafton Central Workshops, 4/1/1971
(j)   ex BR, Thornaby, Cleveland, 19/9/1972
(k)   ex North Gawber Colliery, West Yorkshire, 30/1/1976

(1)   scrapped, after 5/1950, by 5/1951
(2)   derelict, 4/1953; scrapped. 5/1953
(3)   scrapped, after 4/1966, by 4/1967
(4)   scrapped on site by Thos.W.Ward Ltd, 6/1972
(5)   scrapped on site by Walter Heslewood Ltd, c12/1973 (after 9/1973)
(6)   scrapped by Marple & Gillott Ltd, /1948
(7)   scrapped on site by C.F.Booth Ltd, 10/1970
(8)   scrapped on site by unknown merchant, of Sheffield, 12/1968
(9)   to South Kirkby Colliery, West Yorkshire, 5/4/1973
(10)   to Park Mill Colliery, West Yorkshire, 27/10/1972
(11)   to C.F. Booth Ltd, Rotherham, 25/3/1986, and scrapped there, after 25/4/1986, by 21/5/1986
(12)   to C.F.Booth Ltd, Rotherham, 26/3/1986, and scrapped there, after 7/4/1986, by 18/4/1986
(13)   to Woolley Colliery, West Yorkshire, 3/1978 (by 9/3/1978)

Gauge : 2ft 1½in (Underground Locomotives)

No.1		0-4-0DMF		HC	DM703	1947	(a)	(1)
-		0-4-0DMF		HC	DM704	1948	(a)	(2)
-		4wBEF		GB	2838	1957	New	(3)
-		4wBEF		GB	2839	1957	New	(4)
TL33		4wBEF		GB	2842	1957	New	(5)
No.1		0-6-0DMF		HC	DM889	1955	New	(6)
No.2	TL30	0-6-0DMF		HC	DM1365	1965	New	(7)
No.2		0-6-0DMF		HC	DM1244	1961	New	(7)
-		0-6-0DMF		HC	DM944	1955	(b)	(8)
-		4wBEF		GB	2908	1959	(c)	(5)
-		4wBEF		CE	B3206	1985	New (d)	
			rep	CE	B3309	1986		
			rep	CE	B3553	1989		(9)
-		4wBEF		CE	B3200A	1985		
			(plated		B3200)		(e)	(9)
-		4wBEF		CE	B3200B	1985		
			(plated		B3201/1)		(f)	(10)
-		4wBEF		CE	B3000A	1983		
			rep	CE	B3118C	1985		
			rep	CE	B3484	1988	(g)	(11)
-		4w4wBEF		CE	B3773A	1991	New	(12)

(a)   ex Frickley Colliery, West Yorkshire, /1953
(b)   ex New Monckton Colliery , West Yorkshire, 9/1968
(c)   ex New Monckton Colliery , West Yorkshire, c/1969
(d)   to CE, Hatton, Derbyshire, /1986; returned, 9/1986;
       to CE, Hatton, Derbyshire, 10/1988; returned, 2/1989
(e)   ex Riddings Colliery, West Yorkshire, /1989;
       to CE, Hatton, Derbyshire, /1990; returned, /1990
(f)   ex Riddings Colliery, West Yorkshire, /1989
(g)   ex Prince of Wales Colliery, West Yorkshire, /1989

(1)   to Darfield Colliery, /1960
(2)   to Parkmill Training Centre, West Yorkshire, 10/1976
(3)   transferred, sold or scrapped, by /1978
(4)   to Houghton Main Colliery, by 9/1968
(5)   abandoned underground, 8/1976
(6)   to Frickley Colliery, West Yorkshire, c/1958
(7)   to Walkden Central Workshops, Greater Manchester, 12/1980

(8)   written off, by 7/1978
(9)   sold, scrapped or abandoned underground, c/1993
(10) to Prince of Wales Colliery, West Yorkshire, /1989
(11) to CE, Hatton, Derbyshire, /1994, and scrapped there, after 22/7/1996, by 18/11/1996
(12) to CE, Hatton, Derbyshire, 11/1993; thence to Kellingley Colliery, North Yorkshire

HANDSWORTH NUNNERY (HIGH HAZELS) COKING PLANT, Handsworth N37

ex **Nunnery Coke & Gas Co Ltd** (subsidiary of **Nunnery Colliery Co Ltd**)   SK 408876
NE(C) from 1/1/1947; CPD from 1/1/1963.   CLOSED 1963
See HANDSWORTH COLLIERY for description of rail connections.
Gauge : 4ft 8½in

-	0-4-0WE		ACEC		1928	(a)	(1)
	0-4-0WE		GB	1348	1934	(b)	(1)
No.9	0-4-0ST	OC	HL	2454	1900	(c)	(2)

(a)   ex Nunnery Colliery Co Ltd, with site, /1/1947
(b)   ex Dalton Main Coking Plant, c/1954
(c)   ex Rotherham Main Coking Plant, 10/1/1963

(1)   sold or scrapped c/1966 (after 10/1965)
(2)   to Rotherham Main Coking Plant, 3/1963 or 4/1963

HANDSWORTH COLLIERY(HIGH HAZELS SCREENS), Handsworth   N38

ex **Nunnery Colliery Co Ltd**   Colliery SK 409876. Screens SK 409870
NE1 from 1/1/1947; SYK 26/3/1967.   CLOSED 10/1967
Sidings on the north side of the BR (ex LNER) line, 1 mile east of Darnall Station, served HIGH HAZELS SCREENS and the adjacent COKING PLANT. A narrow gauge surface, rope-worked, tubway, which crossed the BR line by an overbridge, linked the screens with HANDSWORTH COLLIERY, ½ mile to the south. Note: It is unclear whether locomotives were owned by colliery or coking plant at any given date.
Gauge : 4ft 8½in

	WAVERLEY	0-4-0ST	OC	AB	889	1901	(a)	(1)
(No.2)		0-6-0ST	IC	HE	786	1902	(a)	(2)
(10)	(No.2)	0-6-0ST	OC	AE	1472	1904		
			reb	C&J		1938	(a)	(3)
No.1		0-6-0ST	IC	HL	3002	1913	(a)	(4)
	HUNTSMAN	0-6-0ST	OC	AB	2018	1936	(b)	(5)
	VICTORY	0-4-0ST	OC	AB	1654	1920	(c)	(6)
	KITCHENER	0-4-0ST	OC	MW	1843	1915	(d)	(7)
	BIRLEY No.5	0-4-0ST	OC	P	1454	1917	(e)	(8)
DL5		0-6-0DM		HC	D1174	1959	New	(9)
	DAISY	0-6-0ST	OC	AE	1895	1923	(f)	(10)

(a)   ex Nunnery Colliery Ltd, with site, 1/1/1947
(b)   ex Brookhouse Colliery, after 18/4/1949, by 1/1/1950;
        to Treeton Colliery, 5/1950 (by 28/5/1950); ex Brookhouse Colliery after 9/1955, by 12/1955
(c)   ex Maltby Colliery, after 28/5/1950, by 2/3/1952
(d)   ex Thos.W.Ward Ltd, dealers, Sheffield, hire, c5/1950 (by 28/5/1950);
        possibly returned off hire, but here again in 4/1951
(e)   ex Maltby Colliery, after 6/7/1952, by 8/1955
(f)   ex Firbeck Colliery, Nottinghamshire, after 8/1964, by 4/1965

(1) to Nunnery Colliery, c/1947
(2) derelict by 18/4/1949; scrapped by Thos.W.Ward Ltd, after 9/1957, by 4/1958
(3) to Orgreave Colliery 4/1962
(4) scrapped by Thos.W.Ward Ltd, after 5/1963, by 10/1963
(5) to Orgreave Colliery, after 8/1964, by 11/1964
(6) to Kiveton Park Colliery, after 3/1954, by 3/1955
(7) to Thos.W.Ward Ltd, Sheffield, off hire, 6/1951;
     later on hire at Manvers Colliery
(8) to Waleswood Coking Plant, 11/1955
(9) to Kiveton Park Colliery, after 8/1964, by 8/1965
(10) to Firbeck Colliery, Nottinghamshire, after 4/1965, by 9/1966

HATFIELD COLLIERY, Stainforth   N39
ex **Hatfield Main Colliery Co Ltd**   SE 654113

NE2 from 1/1/1947; DCR from 26/3/1967; SYK 1/10/1985;
SYG from 1/4/1990; CCG from 1/9/1993; care & maintenance from 3/12/1993.

                                                 sold to **Coal Investments Ltd** 7/1994.

Served by extensive sidings on the north-west side of the BR (ex LNER) line, ¾ miles north east of
Stainforth & Hatfield Station. A line ran north to a staith on the Stainforth & Keadby Canal (¾ mile) and
en route served extensive dirt tips. Use of standard gauge locomotives ceased 1981. Locomotives
were used extensively underground.

Gauge : 4ft 8½in
In addition to the locomotives listed below, 200HP 0-6-0DM locomotives were hired from British
Railways in 1967. These BR locomotives were frequently changed.

	HATFIELD No.3	0-6-0ST	IC	K	3819	1899	(a)		(1)
	HATFIELD No.4	0-6-0T	IC	AE	1448	1902	(a)		(2)
	HATFIELD No.1	0-6-0ST	OC	HL	3197	1916	(a)		(3)
	HATFIELD No.5	0-6-0T	IC	NB	21520	1917	(a)		(4)
	HATFIELD No.2	0-6-0ST	OC	YE	1787	1922	(a)		(5)
6	(71) (HATFIELD No.6)	0-6-0ST	OC	HC	1349	1918			
		reb		YE		1946	(b)		(6)
	THORNE No.1	0-6-0ST	IC	HE	3714	1951	(c)		(7)
HAM 955	3219/015								
	HATFIELD No.7	0-6-0DM		HC	D955	1955	New		(8)
	-	0-6-0DM		HE	5240	1957	New		(9)
39	HICKLETON MAIN No.4	0-6-0ST	IC	HE	3713	1951	(d)		(10)
RM2020	3219/020	390/MM/M/5494							
	TOMMY	0-4-0DH		HC	D1386	1966	(e)		(11)
D2518	3219/016	0-6-0DM		HC	D1209	1961	(f)		(12)
D2519	3219/017 HAM 280	0-6-0DM		HC	D1210	1961	(g)		(13)
D2616		0-6-0DM		HE	5665	1960	(h)		(14)

(a) ex Hatfield Main Colliery Co Ltd, with site, 1/1/1947
(b) ex Appleby-Frodingham Steel Co Ltd, Scunthorpe, Lincolnshire, 12/1950
(c) ex Thorne Colliery, after 5/1956, by 4/1957
(d) ex Hickleton Colliery, after 5/1966, by 6/1967
(e) ex Askern Colliery, after 28/7/1980, by 1/1981
(f) ex BR, Crewe, Cheshire, 8/1967
(g) ex BR, Crewe, Cheshire, 2/1968
(h) ex BR, Goole, East Yorkshire, 5/1968

(1) scrapped on site, c21/3 /1963
(2) scrapped by Raynor Contractors Ltd, of Sheffield, after 10/1967, by 6/1968

(3) dismantled by 8/1967; scrapped on site by Raynor Contractors Ltd, 4/1969
(4) scrapped, after 4/1967, by 9/1957
(5) to Bullcroft Colliery 3/1957
(6) scrapped on site by Raynor Contractors Ltd, of Sheffield, 4/1969
(7) to Thorne Colliery after 10/1967, by 4/1968
(8) sold to R.E. Trem Ltd, c6/1981; engine removed and remains scrapped on site by 10/1981
(9) to Yorkshire Main Colliery, after 4/1958, by 7/1961
(10) scrapped on site by Raynor Contractors Ltd, of Sheffield, 4/1969
(11) to C.F. Booth Ltd, Rotherham, 13/3/1986 and scrapped there, 3/1986
(12) written off, 3/5/1974; scrapped on site by NCB after 5/1974, by 4/1976
(13) to Keighley & Worth Valley Railway, Haworth, West Yorkshire, 3/4/1982
(14) dismantled 4/1973; scrapped by 12/1973

Gauge : 2ft 0in (Surface stockyard)

-		4wDM	RH	249557	1947	(a)	(1)
-		4wDM	RH	189958	1938	(b)	(2)
-		4wDM	RH	249559	1947	(c)	(1)
-		4wDM	RH	221590	1943	(d)	Scr c/1974
No.681	No.1 390/HA/M/0689	0-4-0DMF	HC	DM689	1948	(e)	(3)
No.682	No.2 390/HA/M/0690	0-4-0DMF	HC	DM690	1948	(f)	(4)
No.683	No.3 390/HF/M/0691	0-4-0DMF	HC	DM691	1948	(g)	Scr c/1980

(a) ex Thorne Colliery, c/1957
(b) ex unknown location, by 7/1960 (originally War Office)
(c) ex Thorne Colliery, after 7/1961, by 6/1965
(d) ex Thorne Colliery, after 9/1968, by 5/1971
(e) ex underground, by 12/1969
(f) ex underground, by 5/1971
(g) ex underground, date unknown

(1) scrapped, c/1972 (after 5/1971)
(2) to Markham Main Colliery, after 6/1965, by 9/1966
(3) scrapped, c/1980 (by 1/1981)
(4) scrapped, c/1983 (after 3/3/1983)

Gauge : 2ft 0in (Underground locomotives)

No.101	No.5 390/HA/M/0630	0-6-0DMF	HC	DM630	1948	New	(1)
No.102	No.6 390/HA/M/0640	0-6-0DMF	HC	DM640	1948	New	s/s
No.103	No.4 390/HA/M/0644	0-6-0DMF	HC	DM644	1948	New	s/s c/1971
No.104	No.7 390/HA/M/0672	0-6-0DMF	HC	DM672	1948	New	s/s c/1971
No.105	No.8 390/HA/M/0673	0-6-0DMF	HC	DM673	1949	New	s/s c/1971
No.106	No.12 390/HA/M/0674	0-6-0DMF	HC	DM674	1949	New	s/s c/1971
No.107	No.10 390/HA/M/0675	0-6-0DMF	HC	DM675	1949	New	(2)
No.108	No.11 390/HA/M/0676	0-6-0DMF	HC	DM676	1949	New	s/s c/1971
No.109	No.9 390/HA/M/0677	0-6-0DMF	HC	DM677	1949	New (a)	s/s c/1971
No.681	No.1 390/HA/M/0689	0-4-0DMF	HC	DM689	1948	New	(4)
No.682	No.2 390/HA/M/0690	0-4-0DMF	HC	DM690	1948	New	(5)
No.683	No.3 390/HF/M/0691	0-4-0DMF	HC	DM691	1948	New	(6)
No.110	No.13 390/HA/M/0716	0-6-0DMF	HC	DM716	1951	New	s/s c/1971
No.111	No.14 390/HA/M/0717	0-6-0DMF	HC	DM717	1949	New (b)	(7)
No.112	No.15 390/HA/M/0718	0-6-0DMF	HC	DM718	1951	New (c)	(8)
No.113	No.16 390/HM/M/0785	0-6-0DMF	HC	DM785	1952	New	s/s c/1971
No.114	No.17 390/HM/M/0786	0-6-0DMF	HC	DM786	1953	New	(7)

No.115	No.18	390/HM/M/0787	0-6-0DMF	HC	DM787	1954	New	s/s c/1971
	No.19	390/HA/M/0932	0-6-0DMF	HC	DM932	1956	New	(7)
	No.20	390/HA/M/0981	0-6-0DMF	HC	DM981	1956	New (d)	(7)
No.101	No.21	390/HA/M/0980	0-6-0DMF	HC	DM980	1955	(e)	(7)
	No.22	390/HA/M/0986	0-6-0DMF	HC	DM986	1956	(f)	(7)
		390/MM/M/2593	0-6-0DMF	HC	DM798	1953	(g)	(7)
		390/MM/M/2569	0-6-0DMF	HC	DM796	1952	(h)	(9)
		390/MM/M/2581	0-6-0DMF	HC	DM797	1953	(j)	(7)
		390/MM/M/2606	0-6-0DMF	HC	DM799	1953	(k)	(7)
		390/T/M/3057	0-6-0DMF	HC	DM1108	1959	(m)	(7)
		-	0-6-0DMF	HE	8844	1980		
				HC	DM1444	1980	(n)	(7)
No.2		524/1962	4wBEF	CE	B1574A	1978	(p)	(7)
No.3		524/1963	4wBEF	CE	B1574D	1978	(p)	(7)
		390/4	0-4-0DMF	HC	DM1080	1957	(q)	(7)
		390/5	0-4-0DMF	HC	DM1167	1959	(q)	(7)
		390/6	0-4-0DMF	HC	DM1168	1959	(q)	(7)
		390/14	0-6-0DMF	HC	DM1410	1969	(q)	(7)
		390/13	0-6-0DMF	HC	DM1411	1969	(q)	(7)
		390/7501	0-6-0DMF	HE	7432	1976		
				HC	DM1425	1976	(q)	(7)
		-	4wBEF	CE	B3249A	1986	(r)	(10)
		-	4wBEF	CE	B3249B	1986	(r)	(11)
		Bo Bo 2	4w-4wBEF	CE	B3602A	1990	(s)	(12)
		-	4w-4wBEF	CE	B3603	1990	New	(7)

(a)   to Carcroft Central Workshops, 22/7/1963; returned, 11/12/1963
(b)   to Carcroft Central Workshops, 10/4/1963; returned, 5/11/1963
(c)   to Allerton Bywater Central Workshops, West Yorkshire, after 4/1968. by 9/1968; returned, 10/1968
(d)   to Ashington Central Workshops, Northumberland, 9/1977; returned, 6/1978
(e)   ex Thorne Colliery, 8/1956;
       to Carcroft Central Workshops, 6/10/1965; returned, 29/12/1966
(f)   ex Thorne Colliery, 8/1956
(g)   ex Markham Main Colliery, 22/7/1963;
       to Allerton Bywater Central Workshops, West Yorkshire, 7/1968; returned, 9/1968
(h)   ex Carcroft Central Workshops, after /1962, by /1966
(j)   ex Markham Main Colliery, 12/3/1968
(k)   ex Markham Main Colliery, by 8/1970
(m)   ex Ashington Central Workshops, Northumberland, 3/1977; earlier Thorne Colliery
(n)   ex Rossington Colliery, 7/1981
(p)   ex CE, Hatton, Derbyshire, 2/1989; earlier Shireoaks Colliery, Nottinghamshire
(q)   ex Cynheidre Colliery, Dyfed, 5/1989 (for spares only; remained on surface)
(r)   ex CE, Hatton, Derbyshire, 3/1993; earlier Houghton Main Colliery
(s)   ex Rossington Colliery, 6/1993

(1)   to Carcroft Central Workshops, c/1966
(2)   to Carcroft Central Workshops, /1964
(3)   to Carcroft Central Workshops, 7/1963
(4)   to surface stockyard system, by 12/1969
(5)   to surface stockyard system, by 5/1971
(6)   to surface stockyard system, date unknown
(7)   to Coal Investments Ltd, with site, 7/1994
(8)   to Askern Colliery, 11/1968
(9)   to Markham Main Colliery, /1973

(10)   to Prince of Wales Colliery, West Yorkshire, /1994 (by 7/1994)
(11)   to Kellingley Colliery, North Yorkshire, /1994
(12)   to CE, Hatton, Derbyshire, for alteration to 2ft 6in gauge, /1994;
       thence to Kellingley Colliery, North Yorkshire

HAZLEHEAD COKING PLANT, Crow Edge

N40

ex **Tinker Bros Ltd**

SE

NE(C) from 1/1/1947.

CLOSED 2/1948

See HAZLEHEAD COLLIERY. Note that this was a very small plant that produced only one wagon
load per day.

HAZLEHEAD COLLIERY, Crow Edge

N41

ex **Tinker Bros Ltd**

SE 181049

NE5 from 1/1/1947.

CLOSED 2/1948

Linked to the main line by the private railway of Hepworth Iron Co Ltd that ran north from BR at
Hazlehead Bridge Station to the colliery and coking plant (1½ miles). The closed colliery probably
became part of the Hepworth Iron Co's SLEDBROOK licensed mine.

HICKLETON COLLIERY, Thurnscoe

N42

ex **Doncaster Amalgamated Collieries Ltd**

SE 465054

NE2 from 1/1/1947; DCR from 26/3/1967; NYK from 1/10/1985;
Merged with GOLDTHORPE 1/1986; NYG from 1/4/1990,   CLOSED by 1/10/1991.

Sidings ran east for ½ mile to the colliery and adjacent brickworks from the BR lines (ex LNER &
LMSR Swinton & Knottingley Joint) line, 2 miles north of Bolton on Dearne and from a short BR
branch which ran north from the BR (ex LMSR) Dearne Valley Railway, south-east of Goldthorpe and
Thurnscoe Station. The second of these connections had closed by 1958, though re-organisation in
1978 to access Goldthorpe Colliery from the Swinton & Knottingley line saw it reinstated. The 3ft 0in
gauge underground system replaced that of 2ft 2in gauge c1960.

Gauge : 4ft 8½in

Note that BR locomotive 08142 may have been on hire at Hickleton in 1984. It was seen at Healey
Mills depot and was said to have been coming here.

	SIR WILLIAM	0-6-0ST	OC	AE	1948	1924	(a)	(1)
31	BEN	0-6-0ST	OC	AE	2069	1935	(a)	(2)
No.33	D.A.C.	0-6-0ST	IC	HE	2081	1940	(a)	(3)
No.6	CHESTERFIELD	0-6-0ST	IC	MW	1667	1906	(b)	(4)
No.34	CLEMENT	0-6-0ST	IC	HE	1983	1940	(c)	(5)
HM7321	3219/018	0-4-0DH		HC	D1340	1966	New (d)	(6)
39	HICKLETON No.4	0-6-0ST	IC	HE	3713	1951	New	(7)
D2599	3219/013	0-6-0DM		HE	5648	1960	(e)	(8)
No.30		0-6-0DM		HE	1724	1934		
		reb to 0-6-0DH		HE		1962	(f)	(9)
	F/SE/353 3219/012	0-4-0DH		HC	D1342	1966	(g)	(10)

(a)   ex Doncaster Amalgamated Collieries Ltd, with site, 1/1/1947
(b)   ex Yorkshire Main Colliery, after /1948, by 9/1949
(c)   ex Brodsworth Colliery, 4/1964
(d)   to HE, Leeds, West Yorkshire, for overhaul, 24/9/1977; returned, 24/7/1978

(e)   ex BR, Goole, East Yorkshire, 5/1968
(f)   ex Brodsworth Colliery, after 1/1969, by 27/4/1969;
       to Bullcroft Colliery, after 6/1969,by 11/1969; ex Brodsworth Colliery, after 4/1972,by 3/1973
(g)   ex Frickley Colliery, West Yorkshire, 15/6/1976

(1)   scrapped after 1/4/1956, by 3/1957
(2)   to Yorkshire Main Colliery after 9/1961, by 5/1963
(3)   scrapped, after 1/7/1969, by 10/1969
(4)   to Yorkshire Main Colliery, after 9/1949, by 5/1951
(5)   to Brodsworth Colliery, 29/5/1964
(6)   to Booth Roe Metals Ltd, Rotherham, 13/1/1988; scrapped, after 13/2/1988, by 17/4/1988
(7)   to Hatfield Colliery, after 29/10/1966, by 6/1967
(8)   to Frickley Colliery, West Yorkshire, after 9/1968, by 1/1969
(9)   scrapped on site, after 25/9/1974, by 7/1975
(10)   out of use, by /1982; scrapped on site by unknown contractor, 10/1987 (by 15/10/1987)

Gauge : 3ft 0in (Surface stockyard)

6	390/H/M/3889	0-6-0DMF	HC	DM1249	1961	(a)	(1)

(a)   ex underground, 15/3/1982

(1)   scrapped on site by unknown contractor, 10/1987 (by 15/10/1987)

Gauge : 3ft 0in (Underground locomotives)

	390/H/M/438	0-4-0DMF	HE	3287	1945	(a)	Scr c/1980
6	390/H/M/3889	0-6-0DMF	HC	DM1249	1961	New	(1)
	390/H/M/3898	0-6-0DMF	HC	DM1250	1961	New	s/s
	390/BR/M/318	0-6-0DMF	HC	DM908	1956	(b)	s/s
	390/BR/M/313	0-6-0DMF	HC	DM904	1956	(c)	s/s

(a)   ex Brodsworth Colliery, by 6/1959
(b)   ex Brodsworth Colliery, 10/11/1964
(c)   ex Brodsworth Colliery, by 10/1975

(1)   to surface stockyard, 15/3/1982

Gauge : 2ft 2in (Underground locomotives)

HM28	0-4-0DMF	HE	3128	1944	(a)	Scr c10/1966
390/H/M/458	0-6-0DMF	HC	DM956	1955	New (b)	(1)
390/H/M/466	0-6-0DMF	HC	DM983	1955	New	s/s
390/H/M/2482	0-6-0DMF	HC	DM1109	1959	New	s/s
390/MM/M/2655	0-6-0DMF	HC	DM979	1956	(c)	s/s

(a)   ex Yorkshire Main Colliery 11/1950
(b)   to Carcroft Central Workshops, c/1965; returned, 21/4/1966
(c)   ex Carcroft Central Workshops, 7/1964

(1)   written off by 7/1978

HICKLETON TRAINING CENTRE, Thurnscoe   N43

SE 468053

Opened by DCR 1968.   CLOSED c1976

Located within the Hickleton colliery complex, this comprised a steeply graded narrow gauge track used for training underground locomotive drivers.

Gauge : 3ft 0in

	390/H/M/7692	0-4-0DMF	HE	4044	1949		
		reb to 4wDHF	#		c1957	(a)	(1)
No.1	390/BR/M/774	0-6-0DMF	HC	DM1120	1958	(b)	(2)
	-	4w-4wDHF	HE	7099	1973	New (c)	(3)

rebuilt at NCB Central Engineering Establishment, Bretby, Derbyshire

(a)   ex Mining Research and Development Establishment, Swadlincote Test Site, Derbyshire, c/1968
(b)   ex Brodsworth Colliery, 6/1968
(c)   to HE, Leeds, West Yorkshire, 9/1973; thence to Easington Colliery, Co.Durham

(1)   to Frickley Colliery, West Yorkshire, for spares, c/1974, after 2/6/1974
(2)   to Bentley Training Centre, 5/1976

HIGHGATE COLLIERY, Goldthorpe   N44

ex **Highgate Colliery (1943) Ltd**   SE 459048

NE2 from 1/1/1947; DCR from 26/3/1967.

Merged with GOLDTHORPE 3/1968 and surface CLOSED

Served by a rail connection to the BR(ex LMSR) Dearne Valley line immediately south of the junctions between HICKLETON COLLIERY sidings and the BR lines. Shunting was by main line locomotive and rail traffic ceased c1968 when coal winding was transferred to GOLDTHORPE. In the early 1960s, there is a report of some work being done here by Hickleton locomotives.

HOLBROOK No.1 & 3 CLOSED COLLIERY   N45

ex **J. & G. Wells Ltd**   SK 443812

EM1 from 1/1/1947.   Production ceased 2/1944

Sidings on the west side of the LMSR line at Killamarsh Station, and also from the LNER Holbrook branch, which ran south from Beighton Junction, alongside the LMSR lines, served the colliery. What use was made of rail facilities after production ceased is not known. The site was managed as part of WESTTHORPE COLLIERY, Derbyshire and maintained as a service shaft.

Gauge : 4ft 8½in

No.2 HAZEL	0-4-0ST	OC	AE	1792	1918	(a)	(1)

(a)   ex J. & G. Wells Ltd, with site, 1/1/1947

(1)   to Hartington Central Workshops, Derbyshire, by 3/1949

HOUGHTON MAIN COKING PLANT, Little Houghton   N46a

ex **Houghton Main Colliery Co Ltd**

NE(C) from 1/1/1947.   CLOSED /1950

See HOUGHTON MAIN COLLIERY.

ex **Houghton Main Colliery Co Ltd**   SE 420060

NE4 from 1/1/1947; NE5 from 26/4/1964; Coal wound at GRIMETHORPE from 1964 to 1979;
BNY from 26/3/1967; Rail traffic resumed 1979;
NYK 1/10/1985; NYG from 1/4/1990; SYG from 1/10/1991.   CLOSED 30/4/1993

The colliery and adjacent coking plant were located on the north-east side of the BR (ex LMS) line, 1 mile north of Darfield and at the terminus of an ex LNER branch line running east from Stairfoot. Also connected to the ex LMS Dearne Valley line which passed to the north-east of the colliery. These lines were replaced in the 1950s by a new connection that ran north and then north-west from the BR (ex LMSR) line, 1½ miles south-east of Cudworth Station, to the colliery and also served Grimethorpe Colliery.

During the second period of rail traffic all regular movement was by main line locomotive. The NCB locomotive was little used. The duties of the HC and CE underground locomotives are unknown. The GB locos shunted at the shaft bottom.

Gauge : 4ft 8½in

No.3	HM16	0-6-0ST	OC	MW	1589	1902		
				reb	YE	1922	(a)	1)
No.5		0-6-0ST	IC	P	1303	1913	(a)	(2)
	VICTORY HM15	0-6-0ST	IC	P	1519	1919	(b)	(3)
	HOUGHTON No.4	0-6-0ST	OC	HL	3899	1936	(a)	(4)
No.2	(75128)	0-6-0ST	IC	HE	3178	1944	(c)	(5)
	DARFIELD No.1	0-6-0ST	IC	HE	3783	1953	(d)	(6)
	-	0-4-0ST	OC	P	2108	1950	(e)	(7)
D2199	ROCKINGHAM COLLIERY No.1							
		0-6-0DM		Sdn		1962	(f)	(8)

(a)   ex Houghton Main Colliery Co Ltd, with site, 1/1/1947
(b)   ex Houghton Main Colliery Co Ltd, with site, 1/1/1947;
       to South Kirkby Colliery, West Yorkshire,, c/1954; returned, 'c/1954'
       (4-5 months after arrival)
       to Dearne Valley Colliery, 28/5/1955; returned, 7/1955
(c)   ex Grimethorpe Colliery, after 7/1947
(d)   ex Darfield Colliery, 15/7/1955
(e)   ex Mitchell Main Colliery, 25/2/1957
(f)   ex Barrow Colliery, after 5/1979, by 9/1979

(1)   scrapped, /1957 (after 7/1957)
(2)   to Shafton Central Workshops 7/1964
(3)   dismantled, by 15/7/1961; sold or scrapped, after 5/1962, by 12/1963
(4)   to Upton Colliery, West Yorkshire, after 12/1963, by 3/1964
(5)   to Grimethorpe Colliery, after 9/9/1951, by 4/1953
(6)   to Darfield Colliery, /1959
(7)   to Upton Colliery, West Yorkshire, after 27/5/1962, by 12/1963
(8)   to Royston Drift Mine, West Yorkshire, 14/8/1980

Gauge : 2ft 2in (Surface stockyard)

TL42	FLYING SCOTSMAN							
	(HOUGHTON MAIN FLYER)	4wDM		HE	6273	1965	(a)	(1)
	-	4wDM		HE	7274	1973	(b)	(2)
	DUFFY'S PUFFER	4wDM		HE	7530	1977	New	(1)

(a)	earlier Wombwell Colliery, possibly then via Dearne Valley Colliery;

(a) earlier Wombwell Colliery, possibly then via Dearne Valley Colliery;
   to here, after 9/1968, by 28/8/1973;
   to Philadelphia Central Workshops, Co.Durham, 18/11/1977; returned, /1978
(b) ex Darfield Colliery, 30/8/1988
(1) to Yorkshire Mining Museum, Caphouse Colliery, West Yorkshire,
   after 3/9/1990, by 19/9/1991
(2) to Yorkshire Mining Museum, Caphouse Colliery, West Yorkshire,
   after 28/5/1991, by 19/9/1991

Gauge : 2ft 2in (Underground locomotives)

-	0-4-0DMF	HE	4076	1950	New	(1)
-	0-4-0DMF	HE	4077	1951	New	(1)
-	4wBEF	GB	2839	1957	(a)	s/s by /1978
-	4wBEF	GB	2846	1958	New	(2)
-	4wBEF	CE	B3249A	1986	(b)	(3)
-	4wBEF	CE	B3249B	1986	(b)	(3)
-	4wBEF	CE	B1574F	1978	(c)	(4)

(a) ex Grimethorpe Colliery, by 9/1968
(b) ex Bullcliffe Wood Colliery, West Yorkshire, by 11/1991;
   to CE, Hatton, Derbyshire, for alteration to 2ft 2in gauge, /1992;
   returned, /1992
(c) ex CE, Hatton, Derbyshire, altered from 2ft 0in gauge, /1992;
   earlier Dinnington Colliery

(1) to Frickley Colliery, West Yorkshire, c/1951
(2) derelict on surface, 3/1964; scrapped, c/1964
(3) to CE, Hatton, Derbyshire, /1993; thence to Hatfield Colliery
(4) disposal unknown – sold, scrapped, transferred or abandoned underground

KENDAL DRIFT, Worsbrough
N47
SE 343038 (approx)

Opened by NE5 1956.
CLOSED 9/1958

This drift mine had no standard gauge rail connection and locomotives were not used underground.

KILNHURST COLLIERY, Kilnhurst
N48

ex **Manvers Main Collieries Ltd**
SK 461967

NE3 from 1/1/1947; SYK from 26/3/1967; Rail traffic ceased 1963.
Merged with MANVERS 1/1/1986; CLOSED 26/2/1988

Colliery sidings were on the east side of the BR (ex LMSR) line and west side of the BR (ex LNER) line, ½ mile south of Kilnhurst Station. Coal was wound at MANVERS COLLIERY from c1956 after which the locomotives handled only dirt. The underground locomotives hauled coal to Manvers.

Gauge : 4ft 8½in

	HALLAMSHIRE	0-6-0ST	OC	AB	1792	1923		
			reb	AB		1938	(a)	(1)
No.41	ELSIE	0-6-0ST	OC	WB	2223	1924		
			reb			1937	(a)	(2)
No.40	TINSLEY	0-6-0ST	OC	AB	2025	1936	(b)	(3)

No.45			0-6-0T	IC	HE	830	1904		
			reb	YE			1931		
			reb	YE			1946	(c)	(4)
No.50			4wVBT	VCG	S	9552	1952	(d)	(5)

(a)   ex Manvers Main Collieries Ltd, with site, 1/1/1947
(b)   ex Manvers Main Collieries Ltd, with site, 1/1/1947;
       to Elsecar Central Workshops, after 8/1953, by 9/1954; returned, after 9/1954, by 11/1954
(c)   ex Wath Colliery, after 8/1953, by 9/1954
(d)   ex Manvers Colliery, after 15/7/1961, by 10/1961

(1)   to Cortonwood Colliery, c/1948 (by 10/1949)
(2)   to Manvers Colliery, after 12/1956, by 4/1957
(3)   to Manvers Colliery, after 8/1961, by 2/1963
(4)   to Wath Colliery, 11/1954
(5)   to Manvers Colliery, after 9/1963, by 11/1963

Gauge : 3ft 0in (Underground locomotives)

No.30	390/14500		0-6-0DMF	HE	4815	1955	(a)	(1)
No.29	390/14501	No.3	0-6-0DMF	HE	4816	1955	(b)	(2)
No.29	390/14506		0-6-0DMF	HE	4817	1955	New	(3)
No.32	390/14508		0-6-0DMF	HE	4818	1955	New (c)	(3)
No.2	390/14509		0-6-0DMF	HE	4819	1956	New	(4)
No.42	390/18820	No.3	0-6-0DMF	HE	4863	1956	New (d)	(4)
No.36	390/6019	No.64	0-6-0DMF	HE	4864	1956	New	(5)
No.63	390/14513	No.5	0-6-0DMF	HE	6654	1966	New	Scr 5/1988
No.9	390/6013	No.1	0-6-0DMF	HE	3516	1948	(e)	(6)
No.53	390/14510	No.4	0-6-0DMF	HE	5607	1960	(e)	(7)
	-		2w-2wBEF	TH	SE102	1977	(f)	(8)
No.44			0-6-0DMF	HE	5314	1956	(g)	(9)
No.46			0-6-0DMF	HE	5316	1956	(h)	(4)

(a)   ex Manvers Colliery (date unknown);,
(b)   ex Manvers Colliery (date unknown); on surface at Kilnhurst, 3/1988,
(c)   to Manvers Colliery 12/1956; returned, by /1988; on surface at Kilnhurst, 2/1988
(d)   to Cadeby Colliery, 13/9/1956; returned, by /1969;
       to Allerton Bywater Central Workshops, West Yorkshire, c/1969; returned c/1972
(e)   ex Cadeby Colliery, c/1972
(f)   ex TH, Kilnhurst (for trials), /1977
(g)   ex Silverwood Colliery, 11/2/1983
(h)   ex Silverwood Colliery, after 16/2/1983, by 7/1983

(1)   to Booth Roe Metals Ltd, Rotherham, 25/2/1989; scrapped, after 18/6/1989, by 11/7/1989
(2)   to Bentley Training Centre, 5/1988
(3)   to Manvers Colliery 21/10/1955
(4)   written off, 2/1988
(5)   to Cadeby Colliery, 11/6/1957
(6)   to Booth Roe Metals Ltd, Rotherham, /1988 (by 17/12/1988);
       scrapped, after 27/1/1989, by 11/3/1989
(7)   to Booth Roe Metals Ltd, Rotherham, /1988 (by 17/12/1988);
       scrapped, after 18/6/1989, by 11/7/1989
(8)   to TH, Kilnhurst, after trials, 1977; thence to Easington Colliery, Co.Durham
(9)   to Manvers Colliery, for spares, 16/1/1984

Gauge : 2ft 0in (Dirt disposal system)

-	4wDM		RH	382808	1955	New	(1)
-	4wDM		MR	9696	1952	(a)	(2)

(a)   ex Manvers Colliery, after 4/1953, by 8/1960

(1)   to Wath Colliery 16/3/1971
(2)   to Cadeby Colliery, after 12/1968, by 8/1970

KIVETON PARK COLLIERY, Kiveton Park   N49
ex **Kiveton Park Coal Co Ltd**   SK 494827

NE1 from 1/1/1947; SYK from 26/3/1967; SYG from 1/4/1990; CCG from 1/9/1993.
CLOSED 30/9/1994

Served by sidings on the south side of the BR (ex LNER) line south of Kiveton Park Station. Also served by an ex LMS branch from Killamarsh until c1958 (by 5/1961). Use of standard gauge locomotives had ceased by 6/1985. Use of underground locomotives not known.

Gauge : 4ft 8½in

	KIVETON No.1	0-4-0ST	OC	MW	1842	1915	(a)		(1)
	KIVETON No.2	0-4-0ST	OC	AB	1650	1919	(a)		(2)
	KIVETON No.3	0-4-0ST	OC	HL	3480	1921	(a)		(3)
	(VICTORY)	0-4-0ST	OC	AB	1654	1920	(b)		(4)
	MANTON No.2	4wVBT	VCG	S	9395	1950	(c)		(5)
	-	0-4-0ST	OC	AB	889	1901	(d)		(6)
	(WALESWOOD No.1)	0-4-0ST	OC	HC	750	1906			
		reb		HC		1930	(e)		(7)
DL5	No.521/1981 No.15	0-6-0DM		HC	D1174	1959	(f)		(8)
(D2209)	TRACEY No.16	0-6-0DM		VF	D210	1953			
				DC	2484	1953	(g)		(9)
D2322	No.24	0-6-0DM		RSHD	8181	1961			
				DC	2703	1961	(h)		(10)
(D2328)	No.31	0-6-0DM		RSHD	8187	1961			
	DINNINGTON No.2			DC	2709	1961	(j)		(11)

(a)   ex Kiveton Park Coal Co Ltd, with site, 1/1/1947
(b)   ex Handsworth Colliery, after 3/1954, by 3/1955;
       to Shireoaks Colliery, Nottinghamshire, after 16/9/1956, by 4/1958;
       returned after 4/1958, by 12/1958
(c)   ex Shireoaks Colliery, Nottinghamshire, after 7/1951, by 4/1952
(d)   ex Firbeck Colliery, Nottinghamshire, by 24/9/1950
(e)   ex Waleswood Coking Plant after 14/1/1961, by 12/5/1961
(f)   ex Handsworth Colliery, after 8/1964, by 6/1965
(g)   ex Manvers Colliery, 13/7/1973
(h)   ex Orgreave Colliery, 29/4/1980
(j)   ex Shireoaks Colliery, Nottinghamshire, 13/5/1982

(1)   dismantled 4/1958; scrapped by 5/1959
(2)   scrapped, after 4/1966, by 9/1966
(3)   scrapped 'c10/1965' (after 6/1965, by 3/1966)
(4)   scrapped, after 4/1966, by 9/1966
(5)   to Steetley Colliery, Nottinghamshire, after 16/9/1956, by 4/1958
(6)   derelict 10/1953; scrapped after 26/6/1955, by 7/1955
(7)   to G. Denton, Staveley, Derbyshire, for preservation, c9/1972 (after 6/1972)

(8)   derelict off track, by 2/1986; scrapped, after 28/2/1986, by 12/12/1986
(9)   scrapped on site by Brinsworth Metals Ltd, 8/1985
(10)   scrapped on site, by unknown Barnsley firm, after 13/8/1985, by 28/11/1985
(11)   to Cortonwood Colliery 18/7/1985

Gauge : 2ft 0in (Surface stockyard)

390/6		0-4-0DMF	HC	DM808	1953	(a)	(1)
390/1		0-4-0DMF	HC	DM663	1952	(b)	(2)
KATY 521/13002		4wDM	MR	40S280	1968	(c)	(3)

(a)   ex underground, c/1978 (by 5/1978)
(b)   ex underground, 2/1986;
        to Shireoaks Colliery, Nottinghamshire, 7/11/1986; returned 19/2/1987
(c)   ex Cadeby Colliery, 15/12/1986

(1)   scrapped, after 30/6/1987, by 9/3/1988
(2)   to underground, 2/1988
(3)   to N. Clayton, Ripon, North Yorkshire, c6/1989

Gauge : 2ft 0in (Underground locomotives)

390/1		0-4-0DMF	HC	DM663	1952	New (a)	(1)	
ML 13	FAITH	0-4-0DMF	HC	DM664	1952	New	(2)	
390/2		0-4-0DMF	HC	DM665	1952	New	(3)	
390/3		0-4-0DMF	HC	DM666	1952	New	(4)	
390/4		0-4-0DMF	HC	DM806	1953	New	(5)	
390/5		0-4-0DMF	HC	DM807	1953	New	(4)	
390/6		0-4-0DMF	HC	DM808	1953	New	(6)	
390/12		0-4-0DMF	HC	DM741	1951	(b)	Scr /1988	
No.5	No.48	0-4-0DMF	HE	4808	1954	(c)	(7)	
No.60	390/12099	0-4-0DMF	HE	5423	1964	(d)	(8)	
No.57	390/12779	0-4-0DMF	HE	5598	1961	(e)	(9)	
No.3		4wBEF	CE	B2259	1980			
			rep CE	B3417	1988	(f)	(10)	
-		4wBEF	CE	B3084	1984	(g)	(10)	
-		4wBEF	CE	B2966A	1982	(h)	s/s	
No.6		4wBEF	CE	B3101A	1984	(j)	(10)	

(a)   to surface stockyard, 2/1986; returned underground, 2/1988
(b)   ex Dinnington Colliery, 7/1964
(c)   ex Cadeby Colliery, /1973 (after 6/1969)
(d)   ex Elsecar Colliery, 1/1984 (but never sent underground)
(e)   ex Elsecar Colliery, 1/1984
(f)   ex Tilmanstone Colliery, Kent, 14/4/1987;
        to CE, Hatton, Derbyshire, 6/1987; returned, 2/1988
(g)   ex Tilmanstone Colliery, Kent, 14/4/1987
(h)   ex CE, Hatton, Derbyshire (converted from 1ft 10 in gauge), 6/1987;
        earlier Treeton Colliery
(j)   ex CE, Hatton, Derbyshire, /1989;
        earlier Shireoaks Colliery, Nottinghamshire

(1)   to Hartwood Exports Ltd, Barnsley, c2/1989
(2)   to Manton Colliery, Nottinghamshire, c/1953

(3)   written off, by 7/1978
(4)   scrapped, by 6/1987
(5)   scrapped, after 8/1985, by 6/1987
(6)   to surface stockyard, c/1978 (by 6/1978)
(7)   on surface by 2/1976; dismantled for spares and remains scrapped on site, 6/1978
(8)   scrapped, /1988 (after 3/1988)
(9)   scrapped, by 6/1987
(10) to Hatfield Coal Co Ltd, Hatfield Colliery, c10/1994

MALTBY COLLIERY, Maltby   N50
ex Amalgamated Denaby Collieries Ltd   SK 550925

NE1 from 1/1/1947; SYK from 26/3/1967; SYG from 1/4/1990; NG from 1/9/1993.
RJB Mining (UK) Ltd from 30/12/1994

Served by sidings which ran south-west from the BR line, ¾ miles north of Maltby Station, to the colliery (½ mile). There were extensive rail served tips west of the colliery. The underground locos were used widely for manriding, coal and supplies haulage.

Gauge: 4ft 8½in

MALTBY No.1	0-4-0ST	OC	HL	3771	1930	(a)	(1)
MALTBY No.2	0-4-0ST	OC	HL	3910	1937	(b)	(2)
BIRLEY No.5	0-4-0ST	OC	P	1454	1917	(c)	(3)
ROTHERVALE No.1	0-6-0ST	OC	YE	2240	1929	(d)	(4)
VICTORY	0-4-0ST	OC	AB	1654	1920	(e)	(5)
MALTBY No.3	0-6-0T	OC	AB	2029	1937	(f)	(6)
ROTHERVALE No.7 PHILIP	0-6-0ST	OC	YE	1021	1909	(g)	(7)
No.1	0-6-0ST	OC	YE	2485	1950	(h)	(7)
14 (68077)	0-6-0ST	IC	AB	2215	1947	(j)	(8)
D2274 (No.1) No.17	0-6-0DM		RSHD	7918	1956		
			DC	2620	1956	(k)	(9)
No.2 (D2335)	0-6-0DM		RSHD	8194	1961		
			DC	2716	1961	(m)	(9)
No.18 (D2248) SUE 2243	0-6-0DM		RSHN	7867	1956		
			DC	2580	1956	(n)	(10)
No.33 (D2334)	0-6-0DM		RSHD	8193	1961		
			DC	2715	1961	(p)	(11)
KEN No.67	0-6-0DH		S	10180	1964	(q)	(12)

(a)   ex Amalgamated Denaby Collieries Ltd, with site, 1/1/1947;
      to Shireoaks Colliery, Nottinghamshire, after 7/1955, by 4/1956;
      ex Dinnington Colliery after 9/1956, by 4/1958
(b)   ex Amalgamated Denaby Collieries Ltd, with site, 1/1/1947;
      to Dinnington Colliery, after 4/1950, by 24/9/1950; returned, after 24/9/1950, by 26/5/1951
(c) ex Brookhouse Colliery, /1948;
      to Brookhouse Colliery, after 18/4/1949, by 19/5/1951; returned after 6/1951, by 6/7/1952
(d)   ex Treeton Colliery, 4/1950 (possibly via Thurcroft Colliery)
(e)   ex Brookhouse Colliery, after 18/4/1949, by 4/1950
(f)   ex Thos.W.Ward Ltd, dealers, Sheffield. after 7/1952, by 7/1953
      earlier River Wear Commissioners, Co.Durham;
      to Thurcroft Colliery after 4/1957. by 4/1958; returned 6/1959
(g)   ex Manton Colliery, Nottinghamshire, after 9/1961, by 2/1963
(h)   ex Samuel Fox & Co Ltd, Stocksbridge, c12/1962 (by 1/1963)
(j)   ex Orgreave Colliery, after 9/1968, by 12/10/1968

(k)   ex BR, Allerton, Merseyside, 6/1969
(m)   ex Manvers Colliery, after 6/1989, by 10/1989
(n)   ex Manvers Colliery, c9/1971 (after 7/1971, by 5/1972)
(p)   ex Dinnington Colliery, 24/2/1986
(q)   ex Cadeby Colliery, 8/5/1987

(1)   scrapped, after 9/1961, by 9/2/1963
(2)   scrapped c9/1960 (after 5/1959, by 5/1961)
(3)   to Handsworth Colliery, after 6/7/1952, by 8/1955
(4)   to unknown dealer for scrap, after 10/1969, by 12/1969
(5)   to Handsworth Colliery, after 22/5/1950, by 2/3/1952
(6)   scrapped, after 9/1961, by 10/1962
(7)   to unknown dealer for scrap, 10/1969
(8)   to Keighley & Worth Valley Railway, Haworth, West Yorkshire, 10/9/1971
(9)   scrapped on site, (work commenced on) 25/8/1980
(10) scrapped on site by Carol & Good Ltd, of Thurcroft, 4/1987
(11) to South Yorkshire Railway Preservation Society, Meadow Hall, Sheffield, 12/11/1988
(12) to South Yorkshire Railway Preservation Society, Meadow Hall, Sheffield, 20-21/5/1990

Gauge : 2ft 6in (Surface stockyard)

No.1		4wBEF		CE	B3434A	1988	New	(1)
No.2		4wBEF		CE	B3434B	1988	New (a)	(1)

(a)   may have also spent a time underground here, working at the pit bottom

(1)   to RJB Mining (UK) Ltd, with site, 30/12/1994

Gauge : 2ft 6in (Underground Locomotives)

No.1	524/101	4wBEF		MV	981	1958	New	(1)
No.2	524/102	4wBEF		MV	982	1958	New	(2)
No.3	524/103	4wBEF		MV	983	1958	New	(3)
No.4	524/104	4wBEF		MV	984	1958	New	(4)
No.5	524/105	4wBEF		MV	985	1958	New	(5)
No.6	524/106	4wBEF		MV	986	1958	New	(3)
No.7	524/107	4wBEF		MV	987	1958	New	(6)
No.8	524/108	4wBEF		Bg	3563	1961		
				MV	1180	1961	New	(7)
No.9	524/109	4wBEF		Bg	3660	1969		
				AEI	1272	1969	New	(8)
No.22	524/110	4wBEF		CE	B1575B	1978	New (a)	(9)
No.23	524/111	4wBEF		CE	B1575C	1978	New	(10)
No.21	524/112	4wBEF		CE	B1575E	1978	New	
			rep	CE	B3245	1986	(b)	(10)
No.8	141/1988	4wBEF		CE	B1575F	1978	New	(11)
No.30	521/114	4wDHF		HE	8954	1979	New	(12)
No.25	524/116	4wBEF		CE	B2238B	1980	New (c)	(10)
No.34		4wDHF		HE	8506	1980	(d)	(13)
No.36		4wDHF		HE	8507	1980	(e)	(10)
-		4wDHF	rack	HE	8505	1981	New	(14)
No.35	524/120	4wDHF		HE	8951	1981	(f)	(12)
No.26	524/121	4wBEF		CE	B2964A	1982	New	(10)
No.27	524/122	4wBEF		CE	B2964B	1982	New	(10)
No.28	524/122	4wBEF		CE	B3142B	1984	New	(10)

No.30	524/67	4wBEF		CE	B3101B	1984	(g)	(10)
No.2		4wBEF		CE	B3434B	1988	New	(10)
No.1		4wBEF		CE	B1574B	1978		
			rep	CE	B3757	1991	(h)	(10)
No.40		4w-4wBEF		CE	B3797	1992	New	(10)

(a) to CE, Hatton, Derbyshire, c3/1984; returned c/1984
(a) to CE, Hatton, Derbyshire, /1985; returned 2/1986
(c) to CE, Hatton, Derbyshire, /1989; returned /1990
(d) ex Ledston Luck Colliery, West Yorkshire, for trials, /1981 (after 22/3/1981)
(e) ex Manvers Training Centre, 11/1/1984
(f) ex Manton Colliery, Nottinghamshire, 14/1/1982
(g) ex CE, Hatton, Derbyshire, 28/3/1991, altered from 2ft 0in gauge
     earlier Shireoaks Colliery, Nottinghamshire
(h) ex CE, Hatton, Derbyshire, /1991, altered from 2ft 0in gauge
     earlier Shireoaks Colliery, Nottinghamshire

(1) out of use from 12/1/1982; written off 26/11/1985 and scrapped, c/1986
(2) written off 26/11/1985 and scrapped, c/1986
(3) scrapped, after 10/1986, by 2/1991
(4) out of use from 14/1/1980; written off 26/11/1985 and scrapped, c/1986
(5) out of use from 24/11/1982; written off 26/11/1985 and scrapped, c/1986
(6) out of use from 11/1/1982; written off 26/11/1985 and scrapped, c/1986
(7) dismantled for spares, /1971 and remains later scrapped
(8) out of use from 24/8/1980; written off 26/11/1985 and scrapped, c/1986
(9) to CE, Hatton, Derbyshire, 8/1984; thence to Shirebrook Colliery, Derbyshire
(10) to RJB Mining (UK) Ltd, with site, 30/12/1994
(11) to CE Hatton, Derbyshire, /1987; thence to Brodsworth Colliery
(12) sold or scrapped, after 25/2/1991, by 18/9/1992
(13) either scrapped by 2/1991, or to RJB Mining (UK) Ltd, with site, 30/12/1994
(14) to Manvers Training Centre, c4/1981

Gauge : 2ft 0in (Underground locomotives)

-	0-4-0DMF	HE	4025	1949	New	(1)
-	0-4-0DMF	HE	4026	1949	New	(2)

(1) to Elsecar Colliery 11/1957
(2) to Valleyfield Colliery, Fife, 4/1961

MANVERS BY PRODUCTS PLANT, Wath on Dearne   N51

SE 453012

Opened by NE(C) c/1955; CPD 1/1/63; Rail traffic ceased c/1980.   CLOSED 7/1985

Located to the east of the MANVERS CENTRAL COAL PREPARATION PLANT, the locomotive was used internally at a Benzol plant until rail traffic ceased. Other shunting was by MANVERS COLLIERY locomotives - which see. Note that the feedstock here was supplied from the coking plant on the opposite side of the BR line.

Gauge : 4ft 8½in

CARBONISATION nO.1	0-6-0F	OC	RSHN	7847	1955	New	(1)

(1) to Manvers Coking Plant, after 4/1973, by 7/6/1975 (it had been there on short-term loan at least once previously, noted 16/5/1970)

MANVERS COKING PLANT, Wath on Dearne N52

ex **Manvers Main Collieries Ltd** SE 452009

NE(C) from 1/1/1947; CPD from 1/1/1963. CLOSED 2/12/1980

The coking plant was located within the MANVERS COLLIERY complex between No 1 and No.2 Collieries, immediately south of the BR (ex LMSR) Sheffield - Leeds line. The NCB took over a battery of 71 ovens worked by the EE locomotive and, in 1955, added a further battery of 91 ovens for which two new GB locomotives were obtained. The original ovens were decommissioned in 1978. We believe that from 1955 one GB loco was used on each battery of ovens. Coke production ceased on 2/12/1980 and demolition followed in 1981.

Gauge : 4ft 8½in (shunting locomotive)

(D2238) CAROL No.17	0-6-0DM		VF	D288	1955		
			DC	2562	1955	(a)	(1)
CARBONISATION No.1	0-6-0F	OC	RSHN	7847	1955	(b)	(2)

Note also that colliery loco DC 2483 was here for short periods on loan;
      11/9/1972 to 26/9/1972; 30/4/1973 to 15/5/1973; 19/7/1973 to 19/9/1973 -
      see Manvers Colliery list for locomotive details.

(a)   ex Manvers Colliery, after 16/5/1970, by 1975;
       to Coventry Homefire Plant, West Midlands, c/1975-1977; returned, c/1975-1977
(b)   ex Manvers By-Products plant (and only infrequently used), after 4/1973, by 7/6/1975
       (it had been here on short term loan at least once previously – noted 16/5/1970)

(1)   scrapped by Ernest Northcliffe & Son Ltd, Parkgate, after 12/3/1982, by 12/1982
(2)   to Hallamshire Railway Society, Penistone, 28/5/1981

Gauge : 4ft 8½in (coke ovens)

-	4wWE	EE	871	1932	New	(1)
-	0-4-0WE	GB	2605	1955	New	(2)
-	0-4-0WE	GB	2606	1955	New	(2)

(1)   We have no observations of this locomotive here in NCB ownership and it may have been disposed of prior to 1955 or on receipt of the 1955 locomotives.
(2)   scrapped on site by Thos.W Ward Ltd, 3/1981

MANVERS CENTRAL WORKSHOPS, Wath on Dearne N53

SE 453007

Opened by NE3 after 1956; SYK from 26/3/1967: HQ from 6/1967. CLOSED

The Central Workshops were at the Manvers No. 1 Colliery site and were not rail connected, however some repairs were carried out by workshops staff in the old MANVERS COLLIERY locoshed.

MANVERS COLLIERY & COAL PREPARATION PLANT, Wath on Dearne   N54

ex **Manvers Main Collieries Ltd**   Colliery SE 448012; Coal Preparation Plant SE449015

NE3 from 1/1/1947; SYK from 26/3/1967.   Colliery CLOSED 3/1988
                                                          Coal Preparation Plant CLOSED 3/1989

The working colliery at vesting day was the MANVERS No.2 or "New" Colliery and was located between the BR (ex LMSR and ex LNER) lines south-east of Wath North (LMSR) and north-east of Wath-on-Dearne (LNER) Station. A connection to the LMSR line was made at the west end of the yard and to the LNER line at Wath Junction, south of the colliery. The original (1938) coke ovens were east of the colliery, on the north-west side of the Wath West Curve. This curve was a BR line and ran north-east from Wath Junction on the LNER, passing underneath the LMSR, to join the LMSR/LNER Joint Swinton & Knottingley line. From Manvers a private railway link ran first west, then north, passing under the LMSR before turning north-east, passing under the Swinton & Knottingley line, to continue to Barnburgh Colliery (which see). Note that by Vesting Day the former Hull & Barnsley line to Wath had closed, though its course was to feature in later developments.

A major reconstruction was completed in 1956 when a Central Coal Preparation Plant was opened. This was located on the north side of the BR (ex LMSR) line. This obliterated the western end of the NCB railway to Barnburgh. Coal from Kilnhurst, Manvers and Wath Collieries was thereafter wound at Manvers and went to the plant by conveyor, which bridged the LMSR line. Barnburgh coal was brought in by its loco over the existing private railway to the Meadow Sidings to the east of the Swinton & Knottingley line. Manvers locos worked the traffic from here to a tippler at the new plant. A large new loco shed was built at the plant and thereafter the old shed at the colliery was used for repairs. The extensive sidings at the new plant were connected, at the west end, to the ex LMSR line near Wath North Station and, at the east end, to the ex LNER Wath Curve.

Tipping of dirt by rail, both from Manvers Colliery and the coal preparation plant, was to be done at WATH COLLIERY (which see) and hence a rail link was needed. The reconstruction works having severed the private line beneath the ex LMSR route, a new connection which made use of the disused Hull & Barnsley alignment, with two reversals en route, was made. Colliery dirt travelled over a new line which ran north-west to join the other flow near the old H&B station. WATH locomotives worked the trains from this point to tips west of that colliery.

In connection with the 1956 track re-modelling a contract was awarded to Eagre Construction Ltd (of Scunthorpe, Lincolnshire) which used its own locomotives on this work, although they were also observed on colliery duties.

New coke ovens were built east of the Wath West Curve, the rail connection being via a bridge over the latter, which was formerly the link to Manvers No.1 Colliery, where coal winding had ceased. When Manvers Colliery ceased production, Barnburgh coal continued to arrive by rail for a further year but there was no outgoing traffic.

Underground locomotives hauled coal to Manvers shaft.

Gauge : 4ft 8½in

In addition to the locomotives listed below, locos were hired from BR, including 68847 from 7/1956 until after 4/1958, and between two and four 350HP 0-6-0DE shunters (exchanged frequently) in the period 1986 – 1989.

Note also that the male Christian names of the locos in this list were those of colliery engineers and were applied from the 1960s.

3 ELSIE	0-6-0ST	OC	HC	285	1889	(a)	(1)
22 (MANVERS MAIN No.6	0-6-0ST	OC	HC	822	1912	(b)	Scr 8/1966
24 (MANVERS MAIN No.7)	0-6-0ST	OC	HC	1077	1914	(a)	(2)
No.33(MANVERS MAIN No.9)	0-6-0ST	IC	P	1578	1921	(a)	(3)
No.36(MANVERS MAIN No.10)	0-6-0T	IC	HC	1531	1924	(a)	(4)
(No.44) (MANVERS MAIN No.12)							
WILF	0-6-0ST	IC	P	1891	1940	(c)	(5)
No.21 (MANVERS No.5)	0-6-0ST	IC	P	1242	1911	(d)	(6)
49 TED (No.15)	0-6-0ST	IC	HE	3701	1950	(e)	(7)

No.	Name	Type		Builder	Works No.	Date	Status	Ref
No.50		4wVBT	VCG	S	9552	1952	New (f)	(8)
No.11	MANVERS MAIN 42	0-6-0T	IC	HC	1690	1937	(g)	(9)
	KITCHENER	0-4-0ST	OC	MW	1843	1915	(h)	(10)
No.4		0-4-0ST	OC	P	1114	1907	(j)	(11)
	DRAKE	0-4-0ST	OC	P	2026	1942	(k)	(12)
No.52	DENNIS	0-6-0ST	IC	HE	3832	1955	New (m)	(13)
51	RAYMOND	0-6-0ST	IC	HE	3834	1955	New	(13)
No.46	No.3	0-6-0ST	OC	AB	1150	1908	(n)	(14)
	RERESBY	0-6-0ST	IC	HC	704	1904	(p)	(15)
45		0-6-0T	IC	HE	830	1904	(q)	(16)
No.27		0-4-0ST	OC	HC	1338	1918	(r)	(17)
No.41	ELSIE	0-6-0ST	OC	WB	2223	1924	(s)	(13)
	-	0-6-0ST	IC	HC	1580	1927	(t)	(18)
48		0-6-0ST	IC	HE	3685	1948	(u)	(19)
39	FREDERICK	0-6-0T	OC	HL	3676	1927	(v)	(13)
No.21	(ARTHUR No.61)							
	(CARL No.61)	0-6-0DM		HC	D1154	1959	New	(20)
	(HAROLD No.57)							
	KERRY No.57	0-6-0DM		RH	347748	1958	(w)	(21)
No.40	TINSLEY	0-6-0ST	OC	AB	2025	1936	(x)	(22)
No.26	(DORIS)	0-6-0ST	OC	AB	1498	1918	(y)	(23)
No.63	(68067)	0-6-0ST	IC	HC	1792	1946	(z)	(19)
65		0-6-0ST	IC	HE	3889	1964	New	(24)
	11	0-6-0ST	OC	YE	1823	1922	(aa)	(25)
10261	RAYMOND 74	0-6-0DH		RR	10261	1966	New	(20)
D2373	DAWN No.1 (JIM)	0-6-0DM		Sdn		1961	(ab)	(26)
	-	0-6-0DH		EEV	D1199	1967	(ac)	(27)
D2208		0-6-0DM		VF	D209	1953		
				DC	2483	1953	(ad)	(28)
D2209	TRACEY (ERNEST)	0-6-0DM		VF	D210	1953		
				DC	2484	1953	(ae)	(29)
D2238	CAROL (TOM)	0-6-0DM		VF	D288	1955		
				DC	2562	1955	(af)	(30)
D2326		0-6-0DM		RSHD	8185	1961		
				DC	2707	1961	(ag)	(31)
(D2334)	No.33	0-6-0DM		RSHD	8193	1961		
				DC	2715	1961	(ah)	(32)
D2335		0-6-0DM		RSHD	8194	1961		
				DC	2716	1961	(ah)	(33)
D2336		0-6-0DM		RSHD	8195	1961		
				DC	2717	1961	(ah)	(34)
(D2337)	DOROTHY No.3	0-6-0DM		RSHD	8196	1961		
				DC	2718	1961	(aj)	(35)
D2213		0-6-0DM		VF	D257	1954		
				DC	2529	1954	(ak)	(36)
D2317	No.10	0-6-0DM		RSHD	8176	1960		
				DC	2698	1960	(am)	(37)
(D2332)	LLOYD	0-6-0DM		RSHD	8191	1961		
				DC	2713	1961	(an)	(38)
D2225	DEBRA	0-6-0DM		VF	D274	1955		
				DC	2548	1955	(ap)	(39)
(D2248)		0-6-0DM		RSHN	7867	1956		
				DC	2580	1956	(aq)	(40)
	GEOFFREY No.60	0-6-0DM		RH	347749	1958	(ar)	(41)
	FRANK No.70	0-6-0DH		RR	10223	1965	(as)	(42)

	ERNEST	0-6-0DH	TH	250V	1974	New	(43)
No.71		0-6-0DH	HE	6286	1965	(at)	(44)
	LESLIE No.68	0-6-0DH	S	10181	1964	(au)	(45)
No.25	FRED 2939	0-6-0DH	YE	2939	1965	(av)	(46)
No.34		0-6-0DH	YE	2913	1965	(aw)	(46)
	3219/027	0-4-0DH	HE	7422	1976	(ax)	(47)
No.2	HARRY No.73	0-6-0DH	HE	6661	1966	(ay)	(48)

(a)   ex Manvers Main Collieries Ltd, with site, 1/1/1947

(b)   ex Manvers Main Collieries Ltd, with site, 1/1/1947;
to Barnburgh Colliery, after 9/1947 by 10/1949;
returned, after 10/1949, by 5/1950

(c)   ex Manvers Main Collieries Ltd, with site, 1/1/1947;
to Elsecar Central Workshops, after 7/1956, by 4/1957; returned, after 4/1957, by 4/1958;
to Cortonwood Colliery after 8/1960. by 3/1961; returned, after 16/7/1961, by 11/1962

(d)   ex Barnburgh Colliery, by 18/4/1949

(e)   ex Barnburgh Colliery after 4/1951, by 6/1952;
to Barnburgh Colliery, after 8/1953, by 29/5/1955; returned after 4/1958, by 7/1958

(f)   to Barnburgh Colliery, /1954; ex Wath Colliery, after 4/1957, by 4/1958;
to Wath Colliery after 7/1958, by 5/1959; returned after 13/5/1961, by 15/7/1961;
to Kilnhurst Colliery, after 15/7/1961, by 10/1961; returned, after 9/1963, by 11/1963

(g)   ex Barnburgh Colliery, 1953 (after 28/8/1950, by 4/1953);
to Barnburgh Colliery, after 4/1953, by 8/1953; returned, c/1955 (after 8/1953, by 7/1955);
to Elsecar Central Workshops, /1956 (by 7/1956); returned, c/1958

(h)   ex Thos.W.Ward Ltd, Sheffield, hire, c/1954 (earlier on hire at Handsworth Colliery)

(j)   ex John Cashmore Ltd, dealers, Great Bridge, West Midlands, hire,
(transferred directly from Aldwarke Main Colliery), 5/7/1955

(k)   ex Wath Colliery (hired from, or on contract for, Eagre Construction Co Ltd, Scunthorpe,
Lincolnshire), 7/1956;
to Shanks & McEwan Ltd, Corby, Northants, 5/1957; returned, /1957

(m)   to Barnburgh Colliery, 5/1961 (by 13/5/1961); returned, 7/1961;
to Barnburgh Colliery, after 7/1963; by 3/1964; returned, after 3/1964, by 11/1964;
to Barnburgh Colliery, 3/1966 or 4/1966; returned, after 4/1966, by 9/1966

(n)   ex Wath Colliery after 3/1955, by 5/1955;
to Wath Colliery after 3/1964, by 5/1964; returned 4/1965

(p)   ex Rossington Colliery, 14/3/1956

(q)   ex Wath Colliery, 3/1956

(r)   ex Elsecar Colliery, 6/1956

(s)   ex Kilnhurst Colliery, after12/1956, by 4/1957;
to Barnburgh Colliery, after 4/1957, by 4/1958; returned, after 7/1959, by 7/1959;
to Barnburgh Colliery, after 11/1959, by 9/1960; returned, after 9/1960; by 3/1961

(t)   ex Eagre Construction Co Ltd, Scunthorpe, Lincolnshire, /1958
(either for use on contract or as a hire loco)

(u)   ex Barnburgh Colliery, after 7/1958, by 3/1959;
to Barnburgh Colliery, 8/1960 or 9/1960; returned c3/1961 (by 12/3/1961)

(v)   ex Barnburgh Colliery, after 8/1959, by 10/1959;
to Wath Colliery 5/1963; returned, by 7/1963;
to Barnburgh Colliery, after 3/1964, by 5/1964; returned, after 5/1964, by 11/1964

(w)   ex Barnburgh Colliery, after 5/1959, by 8/1960;
to Barnburgh Colliery, 17/6/1974; returned, 24/6/1974;
to Barnburgh Colliery, 5/10/1976; returned, 18/10/1978

(x)   ex Kilnhurst Colliery, after 8/1961, by 2/1963

(y)   ex Cortonwood Colliery, after 7/1961, by 11/1962

(z)   ex BR, Langwith Junction, Derbyshire, 2/1963;
to Elsecar Central Workshops, after 11/1963, by 3/1964; returned, after 10/1965, by 4/1966

(aa) ex Wath Colliery, 4/1966;
        to Wath Colliery, 3/1967; returned, after 9/4/1967, by 8/1967
(ab) ex BR, Bolton, Lancashire, 9/1968;
        to Barnburgh Colliery, 7/1971; returned, 22/10/1975;
        to Barnburgh Colliery, after 10/1975, by 3/1976; returned, by 4/1976;
        to Wath Colliery, 2/1977 or 3/1977; returned, after 7/1977, by 24/9/1977;
        to Wath Colliery, after 5/1978, by 8/1978; returned 11/1978
(ac) ex EEV, Newton le Willows, Lancashire (demonstration) , 12/1967
(ad) ex BR, Crewe, Cheshire, 11/1968
(ae) ex BR, Allerton, Merseyside, 11/1968;
        (note that there were three short loans to Manvers Coking Plant,
        11/9/1972 to 20/9/1972; 30/4/1973 to 15/5/1973; 10/7/1973 to 19/9/1973)
(af) ex BR, Birkenhead, Merseyside, 11/1968
(ag) ex BR, Gateshead, Co.Durham, 2/1969
(ah) ex BR, Darlington, Co.Durham, 6/1969
(aj) ex BR, Darlington, Co.Durham, 6/1969;
        to Barnburgh Colliery, 6/1974; returned, 2/1977
(ak) ex BR, Birkenhead, Merseyside, 9/1969
(am) ex BR, Gateshead, Co.Durham, 12/1969
(an) ex BR, Gateshead, Co.Durham, 1/1970
(ap) ex BR, Wigan, Lancashire, 1/1970;
        to Wath Colliery, 24/6/1974; returned by 7/1975;
        to Barnburgh Colliery, 16/3/1976; returned, by 6/1976
(aq) ex BR, Bradford, West Yorkshire, 6/1970
(ar) ex Barnburgh Colliery, after 16/5/1970, by 7/1971;
        to Barnburgh Colliery, 4/4/1973; returned, 9/7/1973;
        to Barnburgh Colliery, by 9/1973; returned, after 6/1974, by 11/1975;
        to Barnburgh Colliery, 20/12/1978; returned, 11/1/1979;
        to Barnburgh Colliery, 15/1/1979; returned, 18/1/1979
(as) ex Cadeby Colliery after 7/1971, by 9/1971
(at) ex Wath Colliery after 1/8/1974, by 11/1976
(au) ex Silverwood Colliery, after 8/1981, by 3/1982
(av) ex Treeton Colliery, 10/3/1982
(aw) ex Treeton Colliery, 14/9/1982
(ax) ex Yorkshire Main Colliery, after 9/1985, by 4/2/1986
(ay) ex Barnburgh Colliery, after 2/1986, by 7/7/1986

(1)   to Elsecar Central Workshops, 11/1947
(2)   scrapped, after 4/1958, by 3/1959
(3)   dismantled, /1968; tank (& boiler ?) reused on P 1891;
        remains scrapped, /1976 (after 2/1976)
(4)   scrapped, c/1957 (after 9/1957)
(5)   scrapped on site, c9/1971 (after 7/1971, by 12/1971)
(6)   dismantled by 4/11/1962; scrapped c7/1964 (after 3/1964)
(7)   scrapped on site, 9/1971
(8)   scrapped on site, c9/1969 (after 7/1969, by 10/1969)
(9)   scrapped c7/1966 (after 4/1966; by 9/1966)
(10) returned to Thos.W.Ward Ltd, off hire, c/1954
(11) returned to John Cashmore Ltd, Great Bridge, West Midlands (off hire), 12/1955
(12) returned to Eagre Construction Co Ltd, Scunthorpe, Lincolnshire, after 4/1958
(13) scrapped on site, after 1/1970, by 4/1970
(14) scrapped, /1966 (after 9/1966)
(15) scrapped by Frank Green & Sons Ltd, of Stairfoot, c9/8/1956
(16) scrapped, c/1959 (after 4/1959)
(17) to Elsecar Colliery, 8/1956

(18) returned, sold or scrapped, after 10/1959
(19) scrapped on site, 9/1971
(20) to Booth Roe Metals Ltd, Rotherham, 8/4/1988; and scrapped, after 12/4/1988, by 18/4/1988
(21) to Barnburgh Colliery, 25/11/1983
(22) scrapped c7/1966 (after 4/1966, by 17/9/1966)
(23) to Cortonwood Colliery, 11/1962 or 12/1962
(24) to Cadley Hill Colliery, Derbyshire, after 7/6/1975, by 18/6/1975
(25) to Wath Colliery, after 10/1969, by 1/1970
(26) scrapped on site by Ernest Northcliffe & Son Ltd, after 12/3/1982, by 3/1983
(27) to EEV, Newton le Willows, Lancashire, after 4/1948, by 9/1968
(28) to Cortonwood Colliery, after 12/1968, by 30/3/1969
(29) to Kiveton Park Colliery, 13/7/1973
(30) to Manvers Coking Plant, after 16/5/1970, by 1975
(31) dismantled for spares, /1971 and remains scrapped on site, /1975 (after 7/1975)
(32) to Thurcroft Colliery, 8/10/1969
(33) to Maltby Colliery, after 6/1969, by 10/1969
(34) dismantled for spares and remains scrapped after 11/1977, by 3/1978
(35) to South Yorkshire Railway. Preservation Society, Attercliffe, Sheffield, 22/2/1988
(36) dismantled by 7/1975; remains scrapped, 2/1978
        (although parts may have remained in 6/1978)
(37) to Cortonwood Colliery, c5/5/1970
(38) to Cadeby Colliery, 28/8/1975
(39) to Wath Colliery, 8/12/1976
(40) to Maltby Colliery, c9/1971 (after 7/1971, by 5/1972)
(41) to Barnburgh Colliery 1/7/1983
(42) to Booth Roe Metals Ltd, Rotherham, 26/1/1989, and scrapped there, by 11/3/1989
(43) to Booth Roe Metals Ltd, Rotherham, 31/3/1988, and scrapped there, 4/1988
(44) scrapped on site by Ernest Northcliffe & Son Ltd, Parkgate, c8/1982 (after 12/3/1982)
(45) to Booth Roe Metals Ltd, Rotherham, after 11/1987, by 9/4/1988,
        and scrapped there, c6/1988 (after 16/5/1988)
(46) to Booth Roe Metals Ltd, Rotherham, 8/4/1988, and scrapped there, 12/4/1988
(47) to Booth Roe Metals Ltd, Rotherham, 8/4/1988, and scrapped there, by 16/5/1988
(48) to TH, Kilnhurst, for repair, 24/7/1986,
        but sold thence for scrap to C. Soar Ltd, Barnsley 10/9/1987

Gauge : 3ft 0in (Underground locomotives)

No.2	390/14502		0-6-0DMF	HE	3417	1947	New	(1)
No.3	390/14503		0-6-0DMF	HE	3418	1947	New	(1)
-			0-6-0DMF	HE	3431	1947	New	(2)
No.18	390/14505		0-6-0DMF	HE	4069	1950	New	(1)
	524/14507		4wBEF	GB	2402	1953	New	(3)
No.8	390/14504		0-6-0DMF	HE	3515	1948	(a)	(1)
No.30	390/14500		0-6-0DMF	HE	4815	1955	New	(4)
No.31	390/14501	No.3	0-6-0DMF	HE	4816	1955	New	(4)
No.29	390/14506		0-6-0DMF	HE	4817	1955	(b)	(1)
No.32	390/14508		0-6-0DMF	HE	4818	1955	(c)	(4)
No.2	390/14509		0-6-0DMF	HE	4819	1956	(d)	(5)
No.54	390/14511		0-6-0DMF	HE	5608	1960	New	(1)
No.59	390/14512		0-6-0DMF	HE	3433	1948	(e)	(1)
No.44			0-6-0DMF	HE	5314	1956	(f)	(6)

(a)   ex Cadeby Colliery, 6/1954
(b)   ex Kilnhurst Colliery, 21/10/1955
(c)   ex Kilnhurst Colliery, 12/1956

(d)   ex Kilnhurst Colliery, 23/2/1956
(e)   ex Upton Colliery, West Yorkshire, /1962 (by 11/10/1962)
(f)   ex Kilnhurst Colliery (for spares only), 16/1/1984

(1)   abandoned underground after 3/1988
(2)   to Barnburgh Colliery, by /1978
(3)   to Frank Green Ltd, Stairfoot, Barnsley, for scrap, /1962
(4)   to Kilnhurst Colliery, by /1988
(5)   to Kilnhurst Colliery, c/1972
(6)   remains scrapped on site, after 3/1988

Gauge : 2ft 2in (surface stockyard)

| - | 4wDM | # | | (a) | (1) |
| - | 4wDM | # | | (a) | (1) |

\#   FH or MR locos (one was possibly FH2224 - see below)

(a)   ex Manvers Main Collieries Ltd, with site, 1/1/1947

(1)   sold or scrapped after 8/1950

Gauge : 2ft 2in (underground – old workings)

| - | 0-4-0DMF | HE | 3316 | 1946 | (a) | (1) |

(a)   ex Manvers Main Collieries Ltd, with site, 1/1/1947

(1)   to Barnburgh Colliery, at unknown date

Gauge : 2ft 0in (Surface stockyard)

-	4wDM	FH	2224	1939	(a)	s/s
-	4wDM	MR	9695	1952	New	(1)
-	4wDM	MR	9696	1952	New	(2)
-	4wDM	MR	9697	1952	New	(3)

(a)   supplied new to Manvers Main Collieries Ltd at "Manvers No.2 Yard" and assumed to be still extant at vesting – may be one of the 2ft 2in gauge locos listed above.

(1)   to Cadeby Colliery, c/1958 (by 9/1958)
(2)   to Kilnhurst Colliery, c/1958 (after 4/1953, by 8/1960)
(3)   to New Stubbin Colliery, c/1958 (by 8/1960)

MANVERS TRAINING CENTRE, Wath on Dearne   N55

SE 454015

Opened by SYK c1968.   CLOSED 1988

Located west of the Manvers Central Coal Preparation Plant, this facility comprised a steeply graded narrow gauge track used for training underground locomotive drivers.

Gauge : 2ft 0in

| No.101 | 390/HA/M/0630 | 0-6-0DMF | HC | DM630 | 1948 | (a) | (1) |
| No.21 | 22 | 0-4-0DMF | HE | 4115 | 1955 | (b) | (2) |

(a)   ex Carcroft Central Workshops c/1967-1968 (by 10/1968);
        to HE, Leeds, West Yorkshire, for repairs, c/1975 (by 12/2/19750); returned
(b)   ex New Stubbin Colliery, 7/1979

(1)   to Shireoaks Colliery, Nottinghamshire, after 16/2/1983, by 2/1984
(2)   to Walkden Central Workshops, Greater Manchester, 25/1/1982

Gauge : 2ft 6in

8505		4wDHF	HE	8505	1981	(a)	(1)
-		4wDHF	HE	8507	1981	(b)	(2)

(a) ex Maltby Colliery, c4/1981
(b) ex Mining Research and Development Establishment,
     Bretby, Derbyshire, c/1981 (by 30/10/1981);
     to HE, Leeds, West Yorkshire, c/1982; returned c/1982 (by 5/1983)

(1) to Bentley Training Centre, after 5/1988, by 10/1988
(2) to Maltby Colliery, 11/1/1984

MARKHAM MAIN COLLIERY, Armthorpe N56
ex Doncaster Amalgamated Collieries Ltd
SE 616046

NE2 from 1/1/1947; DCR from 26/3/1967; SYK from 1/10/1985; SYG from 1/4/1990;
Care & maintenance from 30/4/1993; CCG from 1/9/1993.   Coal Investments Ltd from 5/1994

Served by sidings which ran north-east from the BR (ex South Yorkshire Joint) line, 2.miles south of
Kirk Sandall Junction, to the colliery (½ mile). Use of standard gauge locomotives ceased 1989. The
underground locomotive system was used for coal haulage prior to 1975

Gauge : 4ft 8½in

	-	0-6-0ST	IC	HE	2688	1943	(a)	(1)
	CHARLES	0-4-0ST	OC	Mkm		1909	(a)	(2)
	HERBERT	0-6-0ST	OC	AE	1950	1924	(a)	(3)
	ARTHUR MM/M/508	0-6-0ST	IC	HE	3782	1952	New	(4)
No.2	TOM	0-6-0ST	IC	HE	1826	1939	(b)	(5)
	HERBERT							
	MM4441 3219/019	0-6-0DH		HE	5590	1964	New	(6)
	TOMMY 390/MM/M/5494							
	RM2020 3219/020	0-4-0DH		HC	D1386	1966	(c)	(7)
	ROBERT 3219/026	0-4-0DH		HE	7405	1974	New (d)	(8)

(a) ex Doncaster Amalgamated Collieries Ltd, with site, 1/1/1947
(b) ex Bullcroft Colliery, after 7/1962, by 2/1963;
     to Bullcroft Colliery, after 2/1963, by 7/1963; returned after 5/1965
(c) ex Rossington Colliery 8/3/1972
(d) to HE, Leeds, West Yorkshire, 27/2/1980; returned, 2/4/1980

(1) scrapped after 7/1972, by 2/1973
(2) scrapped by Eagre Construction Ltd, c1/1967 (after 10/1966)
(3) to Rossington Colliery, after 9/1960, by 3/1961
(4) written off, 7/6/1976; to Titanic Salvage Co Ltd, Ellastone, Staffordshire, 21/1/1977
(5) to Bullcroft Colliery after 7/1967, by 6/1968
(6) to British Coal Corporation, West Drayton Forward Landsale Depot, Middlesex, 14/4/1989
(7) to Askern Colliery, 2/10/1974
(8) to Marple & Gillott Ltd, Sheffield, for scrap, 11/1989

Gauge : 2ft 0in (surface stockyard)

-	4wDM	RH	189958	1938	(a)	(1)
3	4wDM	RH	222066	1943	(b)	(2)
-	4wDM	HE	3550	1949	(c)	(3)

-	4wDM	HE	3551	1949	(c)	(4)
390/MM/M/2533	0-4-0DMF	HC	DM750	1949	(d)	(5)

(a) ex Hatfield Colliery, after 6/1965, by 9/1966
(b) ex unknown location (new to Ministry of Supply)
(c) ex Thorne Colliery, after 7/1961, by 3/1964
(d) ex underground, c/1973 (by 6/1974)

(1) dismantled 9/1971; sold or scrapped, c/1972
(2) to Rossington Colliery, after 10/1964, by 9/1966
(3) sold or scrapped, c/1972 (by 2/1972)
(4) scrapped, /1973 (after 2/1973)
(5) to Moorside Mining Co Ltd, Mosborough, Sheffield, /1977 (after 24/4/1977)

Gauge : 2ft 0in (Underground locomotives)

-	0-4-0DMF	HE	3488	1946	(a)	(1)
-	0-4-0DMF	HE	3519	1947	(b)	(2)
-	0-4-0DMF	HE	3520	1947	(b)	(2)
-	4wDMF	RH	249557	1947	New	(3)
-	4wDMF	RH	249559	1947	New	(4)
-	4wDMF	RH	249561	1947	New	(3)
-	4wDMF	RH	249563	1947	New	(5)
390/MM/M/2618	0-4-0DMF	HE	3608	1948	New	(6)
390/MM/M/2630	0-4-0DMF	HE	3609	1948	New	(6)
390/MM/M/2643	0-4-0DMF	HE	3610	1948	New	(6)
390/MM/M/2521	0-4-0DMF	HC	DM749	1949	New	(7)
390/MM/M/2533	0-4-0DMF	HC	DM750	1949	New	(8)
390/MM/M/2547	0-4-0DMF	HC	DM751	1949	New	(6)
390MM/M/2557	0-4-0DMF	HC	DM752	1949	New (c)	(9)
-	0-4-0DMF	HC	DM795	1952	New	(10)
390/MM/M/2569	0-6-0DMF	HC	DM796	1952	New (d)	(6)
390/MM/M/2581	0-6-0DMF	HC	DM797	1953	New	(11)
390/MM/M/2593	0-6-0DMF	HC	DM798	1953	New (e)	(12)
390/MM/M/2606	0-6-0DMF	HC	DM799	1953	New	(13)
390/MM/M/2655	0-6-0DMF	HC	DM979	1956	New	(14)
390/MM/M/2504	0-6-0DMF	HC	DM1092	1957	New	(6)
390/MM/M/2668	0-6-0DMF	HC	DM1126	1958	New	(6)
390/T/M/3050	0-6-0DMF	HC	DM928	1955	(f)	(6)
-	4wBEF	CE	B3563A	1989	New	(6)
-	4wBEF	CE	B3563B	1989	New	(15)
-	4w4wBEF	CE	B3794B	1991	New	(16)

(a) ex Yorkshire Main Colliery, by /1954
(b) ex West Midlands Division, No.1 Area
(ordered for Deep Colliery, Staffordshire, but probably unused), c/1948
(c) to Carcroft Central Workshops 16/12/1963; returned, 8/3/1966
(d) to Carcroft Central Workshops, 2/1/1962; ex Hatfield Colliery, c/1973
(e) to Carcroft Central Workshops, 7/2/1963; returned, 26/4/1963
(f) ex Carcroft Central Workshops, 4/1/1960

(1) to Rossington Colliery, 10/1954
(2) to Rossington Colliery, 4/1953
(3) to Thorne Colliery, c/1950

(4) to Yorkshire Main Colliery, c3/1951
(5) to Upton Colliery, c/1949
(6) to Coal Investments Ltd, with site, 5/1994
(7) to Carcroft Central Workshops, 5/12/1962; thence to Askern Colliery
(8) to surface stockyard, c/6/1973 (by 6/1974)
(9) to Rossington Colliery, 19/3/1966
(10) to Rossington Colliery, 10/1952
(11) to Hatfield Colliery, 12/3/1968
(12) to Hatfield Colliery, 22/7/1963
(13) to Hatfield Colliery, by 8/1970
(14) to Carcroft Central Workshops, 12/1963
(15) to CE Hatton, Derbyshire, c7/1993;
      altered to 3ft 0in gauge and thence to Silverwood Colliery
(16) to CE Hatton, Derbyshire, by 7/1995;
      thence to RJB Mining Ltd, Rossington Colliery

Gauge : 500mm (Mineranger trapped rail system)

007		2adDHF	UMM	23.07	1970	New	(1)
M/M5191		2adDHF	UMM	40.002	1971	New	(2)

(1) scrapped, c/1974 (but locomotive seen on the bank 7/1975 may have been this one)
(2) to Sharlston Colliery, West Yorkshire, 5/1974

MARKHAM MAIN ROOMHEAT PLANT, Armthorpe   N57
National Coal Board
Opened by CPD c/1968.   CLOSED c/1970
On MARKHAM MAIN COLLIERY site. Not known if rail connected.

MITCHELL MAIN COKING PLANT, Darfield   N59
ex **Mitchell Main Colliery Co Ltd**   SE
NE(C) from 1/1/1947.   CLOSED 25/11/1956
See MITCHELL MAIN COLLIERY

MITCHELL MAIN COLLIERY, Darfield   N60
ex **Mitchell Main Colliery Co Ltd**   SE 393041
NE4 from 1/1/1947.   CLOSED 27/5/1956
Sidings on the south side of the BR (ex LNER) line, I mile north-west of Wombwell Station, served the colliery and adjacent coking plant.

Gauge : 4ft 8½in

Note that 0-4-0ST OC YE 261 1876 has been claimed here in NCB days. However in-depth investigation of old correspondence suggests that this locomotive did not survive at Vesting Day and that all reports in fact referred to YE 832 (which carried the distinctive ogee tank, whilst YE 261 did not)

'No.3'	0-4-0ST	OC	YE	832	1905	(a)	(1)
MITCHELL MAIN No.905	0-4-0ST	OC	HC	905	1910		
		reb	R.R. Paton			(a)	(2)
DARFIELD MAIN No.992	0-4-0ST	OC	P	992	1905	(b)	(3)

| | | - | 0-4-0ST | OC | P | 2108 | 1950 | (c) | (4) |

(a) ex Mitchell Main Colliery Co Ltd, with site, 1/1/1947
(b) ex Darfield Colliery, 6/5/1951
(c) ex Darfield Colliery, 17/7/1954

(1) to Thos.C.Wild Ltd, Sheffield, 25/4/1952

(2) to Thos W.Ward Ltd, for scrap, 6/1954
(3) scrapped on site by Frank Green & Sons Ltd, c12/10/1957
(4) to Houghton Main Colliery, 25/21957

MONK BRETTON COLLIERY, Monk Bretton   N61
ex **Barrow Barnsley Main Collieries Ltd**

SE 372083

NE5 from 1/1/1947; BNY from 26/3/1967.

CLOSED 4/1968

Served by sidings on the east side of the BR (ex LMS) line south of Monk Bretton Station. Locomotives were not used underground.

Gauge : 4ft 8½in

-	0-6-0ST	IC	P	497	1890	(a)	Scr 7/1953
PRIOR	0-4-0ST	OC	K	3881	1899	(b)	(1)
RAYMOND	0-4-0ST	OC	HC	810	1907	(c)	(2)
SENTINEL No.3	4wVBT	VCG	S	9394	1950	New	(3)
(YORK No.1)	0-4-0ST	OC	YE	2474	1949	(d)	(4)
YORK No.2	0-4-0ST	OC	YE	2473	1949	(e)	(5)
H.C.No.1	0-4-0ST	OC	HC	1889	1960	New	(6)

(a) ex Barrow Barnsley Main Colliery Ltd, with site, 1/1/1947
(b) ex Barrow Barnsley Main Colliery Ltd, with site, 1/1/1947;
     to Birdwell Central Workshops, after 4/1953, by 7/1953; returned 20/8/1955
(c) (a BR Movement Order suggests that this loco moved here from MOTTRAM WOOD
     CLOSED COLLIERY (which see), 11/1947)
(d) ex Wombwell Colliery, after 24/12/1950, by 7/1951;
     to Birdwell Central Workshops, after 6/1963, by 3/1964; returned, after 5/1965. by 2/1966;
     to Barnsley Main Colliery, 2/1966; returned, after 2/1966, by 5/1966
(e) ex Barrow Colliery, after 9/1953, by 16/5/1954

(1) scrapped on site by Thos.W.Ward Ltd, 6/1961
(2) later at Thorncliffe Central Workshops, by 4/1948 (and did not return here)
(3) to Barrow Colliery, 24/12/1950
(4) to Wharncliffe Woodmoor Nos. 4-5 Colliery, 30/10/1968
(5) to Barrow Colliery, 19/1/1955
(6) to Dodworth Colliery, 2/10/1968

MOTTRAM WOOD CLOSED COLLIERY, Barnsley   N62
The Barnsley Syndicate

SE 355073

Colliery closed about 1944 and there is no evidence that it was vested in the NCB. However since a locomotive remained here until 11/1947 we include the entry here in case the site was vested. The sidings were located on the east side of the LNER line south of Old Mill Lane Goods Station, Barnsley.

Gauge : 4ft 8½in

RAYMOND	0-4-0ST	OC	HC	810	1907	(a)	(1)

(a)   owned by The Barnsley Syndicate at vesting day, 1/1/1947

(1)   BR Movement Order claims to Monk Bretton Colliery, 11/1947 (unconfirmed) but the
      locomotive was at Thorncliffe Central Workshops, by 4/1948

NEW STUBBIN COLLIERY, Rawmarsh N63
ex **Earl Fitzwilliam's Collieries Co**

NE3 from 1/1/1947; SYK from 26/3/1967.

SK 427967

CLOSED 7/1978

The colliery was located at the end of an NCB line which ran north from the BR (ex LMSR) line, ¾ miles
south-west of Parkgate & Rawmarsh Station (2 miles), and from the BR (ex LNER) line immediately
west of Rotherham Road Station (2½ miles). There were loco sheds at the colliery and ("the Basin
Shed") at Parkgate (SE434950) where there was a staith on the Parkgate branch of the Sheffield &
South Yorkshire Navigation. Also at Parkgate there were connections to the coking plant of South
Yorkshire Chemicals Ltd and to Park Gate Iron & Steel Co Ltd. Underground locomotives were used
for coal haulage from 1955.

Gauge : 4ft 8½in

No.2	(No.1 VICTORIA)	0-4-0ST	OC	YE	118	1869	(a)	(1)
No.1	(No.3 WENTWORTH)	0-4-0ST	OC	YE	120	1869	(b)	(2)
	No.8 SUCCESS	0-6-0ST	OC	FW	382	1878	(c)	(3)
	No.11	0-6-0PT	OC	Bwn	46489	1917	(d)	(4)
No.34	(No.12)	0-6-0T	OC	HC	1523	1925	(d)	(5)
No.32	(ELSIE)	0-6-0ST	OC	HC	285	1889	(e)	(6)
No.37	(No.11)	0-6-0ST	OC	HC	1368	1920	(f)	(7)
	-	4wVBT	VCG	S	9398	1950	(g)	(8)
	ATLAS No.6	0-4-0ST	OC	YE	478	1892	(h)	(9)
No.12		0-4-0ST	OC	Mkm		1914	(j)	(10)
No.23	CLARRIE No.54	0-6-0DM		HC	D1090	1958	New	(11)
	DAVID No.58	0-6-0DM		HC	D1128	1958	(k)	(12)
No.3	(No.2 MILTON)	0-4-0ST	OC	YE	119	1869	(m)	(13)
	-	4wDH		S	10003	1959	(n)	(14)
D2322	No.25	0-6-0DM		RSHD	8181	1961		
				DC	2703	1961	(p)	(15)

(a)   ex Earl Fitzwilliam's Collieries Co, with site, 1/1/1947;
      to Rotherham Main Colliery, after 8/1950, by 12/1950; returned after 12/1950, by 5/1952
(b)   ex Earl Fitzwilliam's Collieries Co, with site, 1/1/1947;
      to Elsecar Colliery, 2/1950; returned after 5/1950, by 8/1950
(c)   ex Earl Fitzwilliam's Collieries Co, with site, 1/1/1947;
      to Elsecar Colliery, 3/1947; ex Elsecar Central Workshops, 9/1948
(d)   ex Earl Fitzwilliam's Collieries Co, with site, 1/1/1947
(e)   ex Wath Colliery, 5/1949
(f)   ex Elsecar Central Workshops, after 27/5/1950, by 9/1950
(g)   ex TH, demonstration, c10/1950
      (moved directly from Wakefield area, possibly NCB Shawcross Colliery)
(h)   ex Rotherham Main Colliery, 5/1956;
      to Aldwarke Main Colliery 3/11/1958; returned 7/1960
(j)   ex Parkgate Iron & Steel Co Ltd, Rawmarsh (hire), 4/1956
(k)   ex Elsecar Colliery, c12/1966 (after 9/1966, by 8/1967)
(m)   ex Aldwarke Main Colliery, after 3/1959, by 9/1959
(n)   ex East Ardsley Colliery, West Yorkshire, on demonstration, c2/7/1959
(p)   ex Orgreave Colliery, after 6/1975, by 8/1975

(1)   scrapped by Frank Green & Sons Ltd, of Stairfoot, 5/1958
(2)   scrapped by Frank Green & Sons Ltd, of Stairfoot, c10/1954
(3)   scrapped, c9/1951 (after 8/1950, by 5/1952)
(4)   scrapped, after 27/5/1950, by 31/5/1952
(5)   to Elsecar Colliery, after 4/1968, by 9/1968
(6)   out of use by 5/1957; scrapped on site by Frank Green & Sons Ltd, of Stairfoot, 7/1958
(7)   scrapped /1966 (after 4/1966, by 8/1967)
(8)   returned to TH, moved directly to demonstration at North Western Gas Board, Warrington, Lancashire, c9/1/1951
(9)   to Elsecar Colliery, after 7/1963, by 7/1964
(10) returned to, off hire, 5/1956
(11) to Dinnington Colliery, 2/1978
(12) to Cadeby Colliery, 23/8/1978
(13) sold or scrapped, after 3/1963, by 5/1963
(14) to Parkgate Iron & Steel Co Ltd, for demonstration, 7/1959
(15) to Orgreave Colliery, after 6/1975, by 8/1975

Gauge : 2ft 0in (Underground locomotives)

No.21		0-4-0DMF	HE	4113	1955	New		(1)
No.22	No.21 390/16201	0-4-0DMF	HE	4114	1955	New		(2)
No.21	(cab) No.22 (radiator)	0-4-0DMF	HE	4115	1955	New		(3)
No.28		0-4-0DMF	HE	4503	1955	New		(2)
No.52		0-4-0DMF	HE	5514	1958	New		(2)
No.55		0-4-0DMF	HE	6059	1962	New	(a)	(4)
No.14	390/12102	0-4-0DMF	HE	3558	1948		(b)	(5)

(a)   to HE, Leeds, West Yorkshire, c/1978; returned, c/1978
(b)   ex Elsecar Colliery, 7/1975

(1)   to Shaw Cross Colliery, West Yorkshire, c/1955
(2)   to Dinnington Colliery, by 4/1979
(3)   to Manvers Training Centre, 7/1979
(4)   to Elsecar Colliery, 7/1979
(5)   to Shireoaks Colliery, Nottinghamshire, 2/8/1979

Gauge : 2ft 0in (surface stockyard)

-	4wDM	MR	7218	1938	(a)	(1)
-	4wDM	MR	9697	1952	(b)	(2)
KATY	4wDM	MR	40S280	1968	New	(3)

(a)   ex Elsecar Central Workshops /1957
(b)   ex Manvers Colliery, c/1958 (by 8/1960)

(1)   sold or scrapped after 5/1958
(2)   to Marple & Gillott Ltd, Sheffield, for scrap, /1980
(3)   to Silverwood Colliery, 2/8/1979

ex **Fountain & Burnley Ltd**

NE6 from 1/1/1947; BNY from 26/3/1967.

Merged with WOOLLEY Colliery 1/1986 & surface CLOSED

An NCB railway ran east from the BR (ex LMSR) line, ½ mile south of Darton Station, to NORTH GAWBER COLLIERY (1 mile). DARTON COLLIERY was located half way along this line on a short branch to the south. After closure in 8/1948 Darton Colliery site was used as a rail served tip. Underground locomotives were not used at this colliery.

Reference : "Industrial Railway Record " No.165, IRS

Gauge : 4ft 8½in

In addition to the locomotives listed below, locos were hired from BR from time to time. For example, ex-LMSR/LYR 0-6-0ST 51447 was here on 19/7/1949 and again on 19/2/1950. LMSR 47572 was here on 3/4/1950.

-		0-6-0ST	OC	RS	3094	1902		
			reb	YE		1934	(a)	(1)
(No.1)		0-6-0T	IC	HC	884	1911		
			reb	YE		1932	(a)	(2)
N3	SALISBURY	0-6-0T	IC	HC	1069	1914	(a)	(3)
71497		0-6-0ST	IC	HC	1774	1944	(b)	(4)
	BRAMLEY No.4	0-6-0ST	OC	HE	1643	1929	(c)	(5)
N2		0-6-0T	OC	HC	1857	1952	New	(6)
N1		0-6-0T	IC	HC	1858	1952	(d)	(7)
-		0-6-0T	OC	HC	1816	1948	New	(8)
		0-6-0ST	OC	HC	614	1902	(e)	(9)
WS5	WATSON	0-6-0ST	OC	HC	1197	1916	(f)	(10)
1353	12	0-6-0ST	IC	YE	2569	1954	(g)	(11)
WD 161		0-6-0ST	IC	HE	3212	1940	(h)	(12)
WD 135	NG/S/153	0-6-0ST	IC	HE	3171	1944	(j)	(13)
	MONCKTON No.1	0-6-0ST	IC	HE	3788	1953		
			rep	HE	59240	1964	(k)	(14)
No.63		0-4-0ST	OC	AB	2195	1945	(m)	(15)
D2284		0-6-0DM		RSHD	8102	1960		
				DC	2661	1960	(n)	(16)
	WOOLLEY No.1	0-6-0DH		AB	553	1969	(p)	(17)
08679		0-6-0DE		Horwich		1959	(q)	(18)
TL13		4wDH		TH	142C	1964		
		built on frame of S					(r)	(18)

(a)   ex Fountain & Burnley Ltd, with site, 1/1/1947
(b)   on loan from MFP (possibly Wath or Skiers Spring) at vesting day
(c)   ex HE, Leeds, West Yorkshire (hire), c/1948
      (but possibly hire was to MFP Darton Disposal Point, not North Gawber Colliery)
(d)   ex Wharncliffe Woodmoor Nos.1-3 Colliery, 8/1954;
      to HE, Leeds, West Yorkshire, for repairs, after 2/1960, by 12/3/1961; returned, by 13/5/1961
(e)   ex Haigh Colliery, West Yorkshire, after /1948, by 3/1950
(f)   ex Wharncliffe Woodmoor Nos.1-3 Colliery, 5/1956
(g)   ex United Steel Companies Ltd, Exton Park, Rutland, for trials, /1961 (by 12/3/1961)
(h)   ex WD, Shoeburyness, Essex, WD 161, 4/1964
(j)   ex WD, Bramley, Hampshire, WD 135, 5/1962;
      to HE, Leeds, West Yorkshire, after 5/1964, by 9/1964; returned, by 3/1965
(k)   ex New Monckton Colliery, West Yorkshire, 14/11/1967
(m)   ex Hartley Bank Colliery, West Yorkshire, 6/1968, (by 23/6/1968)
(n)   ex BR, Colchester, Essex, 16/7/1971
(p)   ex Woolley Colliery, West Yorkshire, 28/1/1974

(q)   ex BR, Allerton, Merseyside, c25/6/1976;
        to Royston Drift Mine, West Yorkshire, 17/9/1976; returned, 5/4/1979
(r)   ex Skiers Spring Colliery, 26/4/1977

(1)   scrapped, by Wm.Hardman, 6/1948 (another source suggests "by 7/1947")
(2)   to HC, Leeds, West Yorkshire, 9/1947 or 11/1947; rebuilt by HC , /1953;
        thence to Wharncliffe Woodmoor Nos.4-5 Colliery
(3)   to Woolley Colliery, West Yorkshire, 9/1962
(4)   either returned to Wath or to Skiers Spring Disposal Point, 14/7/1948
(5)   returned to HE, c/1950
(6)   to Shackerstone Railway Society, Leicestershire, 6/8/1975
(7)   scrapped on site by George Cohen, Sons & Co Ltd, 17-26/3/1971
(8)   to, 8/1954
(9)   to Woolley Colliery, West Yorkshire, 22/6/1949
(10) to Woolley Colliery, West Yorkshire, 2/1962
(11) returned to United Steel Companies Ltd, 6/1961
(12) scrapped on site by Walter Heselwood Ltd, 4/1977
(13) scrapped on site by Cooper (of Chesterfield ?), 9/1971
(14) to Yorkshire Dales Railway, Embsay, North Yorkshire, 23/2/1980
(15) scrapped on site by Walter Heselwood Ltd, 6/1972
(16) to Grimethorpe Colliery, 30/1/1986
(17) to C.F. Booth Ltd, Rotherham, 16/4/1986, and scrapped there by 24/5/1986
(18) to C.F. Booth Ltd, Rotherham, 18/4/1986, and scrapped there by 24/5/1986

NUNNERY BRICKWORKS, Sheffield   N65
ex **Nunnery Colliery Co Ltd**
NE(B) from 1/1/1947.   CLOSED c2/1959
Located at the east end of NUNNERY COLLIERY, which worked any standard gauge traffic. A short
narrow gauge locomotive worked line ran south into a clay hole. This line closed c6/1958.
Gauge : 2ft 0in

-	4wDM	RH	211650	1942	(a)	(1)
-	4wDM	MR			(b)	(2)
-	4wDM	FH	2623	1942	(c)	(2)

(a)   ex Nunnery Colliery Co Ltd, with site, 1/1/1947
(b)   identity unknown; ex Thos.W.Ward Ltd, dealers, Sheffield, hire, /1955
(c)   ex Thos.W.Ward Ltd, dealers, Sheffield, hire, /1955; originally WD

(1)   sold or scrapped after 4/1958
(2)   returned to Thos.W.Ward Ltd, off hire, /1955

NUNNERY COKING PLANT, Sheffield   N66
ex **Nunnery Colliery Co Ltd**
NE(C) from 1/1/1947.   CLOSED 6/1957
See NUNNERY COLLIERY.

NUNNERY COLLIERY, Sheffield   N67

ex **Nunnery Colliery Co Ltd**   SK 377878

NE1 from 1/1/1947.   Colliery CLOSED 28/8/1953; Washery CLOSED 13/10/1961

Located south of the BR (ex LNER) line, 1¼ miles east of Sheffield Victoria Station. A BR (ex LMSR) branch ran east to the colliery from a point north of Sheffield Midland Station and an NCB line ran west for 1 mile to City Goods Station and a connection to other BR lines. It also served the terminal basin of the Sheffield & South Yorkshire Navigation. Another line, which used rope haulage, ran north from the colliery, passing beneath the BR (ex LNER) in a tunnel, beyond which it crossed roads, the canal and the River Don, to reach the coking plant – in later years this location was used as a landsale depot.

Gauge : 4ft 8½in

(No.8)	0-6-0T	OC	HL	2879	1911	(a)	(1)	
(No.9)	0-6-0T	OC	KS	4080	1919	(a)	(1)	
NUNNERY COLLIERY Co Ltd No 7	0-6-0ST	IC	HL	3726	1928	(a)	(2)	
WAVERLEY	0-4-0ST	OC	AB	889	1901	(b)	(3)	
MALTBY No.1	0-4-0ST	OC	HL	3771	1930	(c)	(4)	

(a)   ex Nunnery Colliery Co Ltd, with site, 1/1/1947
(b)   ex Handsworth Colliery, c/1947
(c)   ex Shireoaks Colliery, Nottinghamshire, 13/4/1956

(1)   scrapped on site by C.F.Booth Ltd, 4/1962
(2)   to Brookhouse Colliery, 2/1954
(3)   to Firbeck Colliery, Nottinghamshire, /1948
(4)   to Dinnington Colliery, 5/9/1956

OLD SILKSTONE COKING PLANT, Barugh   N68

ex **Old Silkstone Collieries Ltd**   SE 318083

NE(C) from 1/1/1947.   CLOSED 7/1958

Located on a short branch which ran south west from the BR (ex LMSR) line, 1 mile south east of Darton Station. Shunting was by main line locomotive. This site should not be confused with the adjacent works of Low Temperature Carbonisation Ltd, which had a loco until 1952.

OLD SILKSTONE COLLIERY - see DODWORTH COLLIERY

ORCHARD ROAD DISPOSAL POINT, Canklow
    – alternative name for ROTHERHAM MAIN DISPOSAL POINT

ORGREAVE COLLIERY, Orgreave   N69

ex **United Steel Companies Ltd**   SK 424870

NE1 from 1/1/1947; SYK from 26/3/1967. Rail traffic ceased by 4/1980
Colliery CLOSED 9/1981; Washery CLOSED by 1990

Located on the south-east side of the BR (ex LNER) Treeton Colliery branch, ½ mile north-east of Orgreaves Junction on the Sheffield – Worksop main line. Also connected to the BR (ex LMSR) line at Treeton Station and to the works of United Coke & Chemicals Ltd, which was sited north of the colliery; this works was also connected directly to the ex LNER (NCB worked) branch. Extensive dirt tips to the south of the colliery were served by both a 4ft 8½in gauge railway and an aerial ropeway. Locomotives were not used underground.

Although the colliery closed in 9/1981, its Coal Preparation Plant was retained to wash Treeton Colliery coal, which came by overland conveyor.

Gauge : 4ft 8½in

In addition to the locos listed below, BR (ex LMSR) 41536 (0-4-0T OC) was on hire here on 18/4/1948. Also 41523 (0-4-0ST IC) and 41530 (0-4-0T OC) were noted in the 1948-1950 period.

No.3	(ROTHERVALE No.3)	0-6-0ST	IC	HC	376	1891	(a)	(1)
	(ROTHERVALE No.4)	0-6-0ST	IC	HC	410	1893	(a)	(1)
	ROTHERVALE No.6	0-6-0ST	IC	HC	565	1900	(a)	(2)
No.8	(ROTHERVALE No.8)	0-6-0T	OC	KS	3075	1917	(a)	(3)
	ROTHERVALE No.9	0-6-0ST	OC	HC	1347	1918	(b)	(4)
	-	0-6-0ST	IC	HE	3134	1944	(a)	(5)
(75058)	No.12	0-6-0ST	IC	RSH	7094	1943	(c)	(4)
	MANTON No.2	4wVBT	VCG	S	9548	1952	(d)	(6)
	ROTHERVALE No.7	0-6-0ST	OC	YE	1021	1909	(e)	(7)
10		0-6-0ST	OC	AE	1472	1904		
				reb	C&J		1938	(f)	(8)
14	(68077)	0-6-0ST	IC	AB	2215	1947	(g)	(9)
No.20		0-6-0T	OC	HC	1731	1942	(h)	(10)
	HUNTSMAN	0-6-0ST	OC	AB	2018	1936	(j)	(11)
No.25		0-6-0DH		YE	2939	1965	New	(12)
No.24	(D2322)	0-6-0DM		RSHD	8181	1961		
				DC	2703	1961	(k)	(13)
D2229	No.5 521/1952	0-6-0DM		VF	D278	1955		
				DC	2552	1955	(m)	(14)

(a)   ex United Steel Companies Ltd, with site, 1/1/1947
(b)   ex United Steel Companies Ltd, with site, 1/1/1947;
        to Treeton Colliery, 4/1959; returned; c24/4/1959;
        to Treeton Colliery, c8/1959 (by 11/1959); returned; after 11/1959, by 4/1960
(c)   ex WD, after hire to Belgian National Railways, 3/1947
(d)   ex Treeton Colliery 9/1959
(e)   ex Treeton Colliery, 3/1960
(f)   ex Handsworth Colliery, 4/1962
(g)   ex BR, Colwick, Nottinghamshire, 29/12/1962
(h)   ex Samuel Fox & Co Ltd, Stocksbridge, 31/8/1962;
        to Treeton Colliery,10/1965; returned, 9/3/1967
(j)   ex Handsworth Colliery, after 8/1964, by 11/1964
(k)   ex BR, Gateshead, Co.Durham, 2/1969;
        to Treeton Colliery, c/1972 (after 6/1969, by 6/1972); returned, after 6/1972, by 4/1973;
        to Treeton Colliery, after 4/1973, by 11/1973; returned, after 11/1973, by 4/1974;
        to New Stubbin Colliery, after 6/1975, by 8/1975; returned, after 6/1975, by 8/1975
(m)   ex Brookhouse Colliery, after 8/1970, by 28/7/1971;
        to Brookhouse Colliery, c1/1972 (by 19/4/1972); returned, after 5/1973, by 11/1973

(1)   scrapped, after 11/1959, by 4/1960
(2)   scrapped, after 2/1963, by 5/1963
(3)   scrapped on site by Thos.W.Ward Ltd, 1/1963
(4)   scrapped, after 6/1965, by 1/1966
(5)   to United Coke & Chemicals Ltd, Orgreave, /1954 (after 6/1953, by 8/1955)
(6)   to Steetley Colliery, Nottinghamshire, after 11/1959, by 25/2/1961
(7)   to Treeton Colliery, 17/3/1961
(8)   scrapped, after 3/1964, by 7/1964
(9)   to Maltby Colliery, after 9/1968, by 12/10/1968
(10)   to Cadeby Colliery, after 6/1969, by 4/1970
(11)   scrapped, after 1/1966, by 4/1966
(12)   to Treeton Colliery, after 7/12/1975, by 2/1976

(13)   to Kiveton Park Colliery, 29/4/1980
(14)   to Brookhouse Colliery, after 11/1973, by 7/1974

REDBROOK COLLIERY, Barnsley   N70
National Coal Board   SE 328079
Old colliery reopened by BNY c/1985; NYK from 1/10/1985.   CLOSED 6/1987

There was no standard gauge rail connection to this short-lived colliery, and locomotives were not used underground.

Gauge: 2ft 3in (Surface stockyard system)

		4wDHF	HE	8831	1978	(a)	(1)
-		4wDHF	HE	8832	1978	(b)	(1)

(a)   ex Parkmill Colliery, West Yorkshire, 8/1985
(b)   ex Parkmill Colliery, West Yorkshire, 3/1986

(1)   to Yorkshire Mining Museum, Caphouse Colliery, West Yorkshire, 5/1988

ROCKINGHAM COLLIERY, Birdwell   N71
ex **N.C. Thorncliffe Collieries Ltd**   SE 325009
NE5 from 1/1/1947; BNY from 26/3/1967.   CLOSED 11/1979

Located on the north-west side of the BR (ex LNER) line, 1 mile north-east of Birdwell & Hoyland Common Station. Also connected to the parallel BR (ex LMSR) line until this closed. From 1955 until 1976, most of the output of this colliery was brought to the surface at a drift mine at SKIERS SPRING (which see). Coal was also delivered to Smithywood Coking Plant by a 4½ mile long aerial ropeway. This was disused from 1962 and was dismantled in early 1965. Shunting at Rockingham was initially by horses, whilst BR locomotives shunted the reception siding until 1974

Gauge : 4ft 8½in

(D2199)   ROCKINGHAM COLLIERY 1

	0-6-0DM	Sdn		1961	(a)	(1)

(a)   ex BR, Doncaster Works, 2/1974

(1)   to Barrow Colliery, after 10/1978, by 4/1979

Gauge : 2ft 0in (surface locomotive, never put to use)

C/DL 4	4wDMF	HE	4467	1953	(a)	(1)

(a)   ex Auchincruive Nos.4-5 Colliery, Ayrshire, after 27/9/1973, by /1975

(1)   sold or scrapped, by 7/5/1979

ROSSINGTON COLLIERY, Rossington   N72
ex **Amalgamated Denaby Collieries Ltd**   SK 600984
NE2 from 1/1/1947; DCR from 26/3/1967: SYK from 1/10/1985; SYG from 1/4/1990;
Care & maintenance from 30/4/1993; CCG from 1/9/1993.   to **RJB Mining (UK) Ltd** 21/3/1994

Located at the end of two short BR branches which ran south from the BR (ex LMSR) Dearne Valley Railway at Loversall Carr Junction and west from the BR (ex LNER) East Coast Main Line north of Rossington Station; in later years only the latter connection survived. No standard gauge locomotive from 3/1972. Rossington was the pioneer user of flameproof underground locomotives, initially for manriding and thereafter for all purposes.

Gauge : 4ft 8½in

	(PUP)	0-4-0ST	OC	HC	524	1899	(a)	(1)
	IRENE	0-6-0ST	IC	HC	655	1903	(a)	(2)
	BLYTHE	0-6-0ST	OC	AE	1894	1922	(a)	(3)
No.50		0-6-0ST	OC	HC	1532	1924	(a)	(4)
	RERESBY	0-6-0ST	IC	HC	704	1904	(b)	(5)
	ROSSINGTON No.1	0-6-0ST	IC	HE	3594	1950	New	(6)
	HERBERT	0-6-0ST	OC	AE	1950	1924	(c)	(7)
D2598	SAM	0-6-0DM		HE	5647	1960	(d)	(8)
RM2020	TOMMY 3219/020	0-4-0DH		HC	D1386	1966	New	(9)

(a)   ex Amalgamated Denaby Collieries Ltd, with site, 1/1/1947
(b)   ex Denaby Colliery, c/1948
(c)   ex Markham Main Colliery, after 9/1960, by 3/1961
(d)   ex BR, Goole, East Yorkshire, 17/5/1968

(1)   scrapped by Marple & Gillott Ltd, /1953
(2)   scrapped, after 9/1960, by 3/1961
(3)   scrapped, after 11/1967, by 11/1968
(4)   to Bullcroft Colliery, after 6/1967 by 6/1968
(5)   to Manvers Colliery, 14/3/1956
(6)   to Askern Colliery, 8/6/1968
(7)   scrapped, after 7/1963, by 3/1964
(8)   to Askern Colliery, 6/1971 or 7/1971
(9)   to Markham Main Colliery, 8/3/1972

Gauge : 2ft 0in (surface stockyard)

(No.1) (2) 3	4wDM	RH	252863	1947	New	(1)
(No.2) (1)	4wDM	RH	252864	1947	New	(2)
3	4wDM	RH	222066	1943	(a)	(3)
-	4wDM	RH	331263	1952	(b)	(4)

(a)   ex Markham Main Colliery, after 10/1964, by 9/1966
(b)   ex Calverton Colliery, Nottinghamshire, 11/1976 (c10/11/1976) (altered from 2ft 6in gauge)

(1)   dismantled, by 3/1983; scrapped, after 5/1983
(2)   to P.J. O'Donald, Sherburn in Elmet, North Yorkshire, by 4/1975
(3)   scrapped, c/1978 (after 11/1976)
(4)   out of use, by /1984; scrapped, after 5/1986

Gauge : 2ft 0in (Underground locomotives)

No.1	390/R/M/2000	0-4-0DMF	HE	2008	1939	(a)	(1)
No.2	390/R/M/2012	0-4-0DMF	HE	2388	1941	(a)	(2)
No.3	390/R/M/2024	0-4-0DMF	HE	3149	1945	(a)	(3)
No.4	390/R/M/2036	0-4-0DMF	HE	3511	1947	(b)	(2)
No.5	390/R/M/2048	0-4-0DMF	HC	DM795	1952	(c)	(2)
No.6	390/R/M/2060	0-4-0DMF	HE	3520	1947	(d)	(2)
No.7	390/R/M/2072	0-4-0DMF	HE	3519	1947	(d)	(4)
No.8	(No.7?) 390/R/M/2084	0-6-0DMF	HC	DM800	1953	New	(5)
No.9	(No.8?) 390/R/M/2096	0-6-0DMF	HC	DM801	1954	New (e)	(6)
No.11	(No.9?) 390/R/M/2108	0-6-0DMF	HC	DM839	1954	(f)	(7)
No.11	390/R/M/2120	0-6-0DMF	HC	DM802	1954	New (g)	(8)
No.12	390/R/M/2132	0-6-0DMF	HC	DM803	1954	New	(9)
No.13	No.12A 390/R/M/2144	0-4-0DMF	HE	3488	1946	(h)	(4)

No.14	390/R/M/2156	0-6-0DMF	HC	DM929	1955	New (j)	(6)
No.15	390/R/M/2168	0-6-0DMF	HC	DM930	1955	New (k)	(10)
No.16	390/R/M/2180	0-6-0DMF	HC	DM933	1956	New (m)	(6)
No.17	390/R/M/2192	0-6-0DMF	HC	DM934	1956	New (n)	(6)
No.18	390/R/M/2204	0-6-0DMF	HC	DM935	1956	New	(11)
No.19	390/R/M/2216	0-6-0DMF	HC	DM936	1956	New	(6)
No.20	390/R/M/2228	0-6-0DMF	HC	DM937	1956	New (p)	(12)
No.21	390/R/M/2241	0-6-0DMF	HC	DM982	1956	New	(2)
No.22	390/R/M/7892	0-6-0DMF	HC	DM1151	1959	(q)	(2)
No.23		0-4-0DMF	HC	DM752	1949	(r)	(13)
T1	39O/T/M/3036	0-6-0DMF	HC	DM840	1954	(s)	(6)
T2	390/T/M/3043	0-6-0DMF	HC	DM841	1954	(t)	(6)
R4		0-6-0DMF	HE	8842	1980		
			HC	DM1442	1980	New	(6)
R6		0-6-0DMF	HE	8843	1980		
			HC	DM1443	1980	New	(6)
-		0-6-0DMF	HE	8844	1980		
			HC	DM1444	1980	New	(14)
Bo Bo 2		4w-4wBEF	CE	B3602A	1990	New	(15)
Bo Bo 1		4w-4wBEF	CE	B3602B	1990	New	(6)
-		4wBEF	CE	B3645A	1990	New	(6)
-		4wBEF	CE	B3645B	1990	New	(6)
-		4wBEF	CE	B1574C	1978		
	rebuilt with new frame		CE	B3864B	1992	(u)	(6)
		4wBEF	CE	B1574G	1978		
	rebuilt with new frame		CE	B3864A	1992	(u)	(6)
No.28		0-6-0DMF	HC	DM1331	1964	(v)	(6)
No.26		0-6-0DMF	HC	DM1393	1967	(v)	(6)

(a)   ex Amalgamated Denaby Collieries Ltd, with site, 1/1/1947
(b)   ex Thorne Colliery, c/1947;
      to HE, Leeds, West Yorkshire, c/1962; returned, 11/4/1962
(c)   ex Markham Main Colliery, 10/1952;
      to HE, Leeds, West Yorkshire, c/1960; returned, 19/8/1960
(d)   ex Markham Main Colliery, 4/1953
(e)   to HC, Leeds, West Yorkshire, (unknown date); returned;
      to Ashington Central Workshops, Northumberland, 28/6/1978; returned, c/1979
(f)   ex Thorne Colliery, by 3/1954
(g)   to HC, Leeds, West Yorkshire, 11/5/1967; returned, 21/7/1967
(h)   ex Markham Main Colliery, 10/1954
(j)   to Carcroft Central Workshops, 11/10/1962; returned, 29/1/1963
      to Ashington Central Workshops, Northumberland, 1/1980; returned, 12/1980
(k)   to Ashington Central Workshops, Northumberland, 6/2/1963; returned, 8/1963;
      to HC, Leeds, West Yorkshire, 5/11/1965; returned, 26/4/1966;
      to HC, Leeds, West Yorkshire, 22/12/1966; returned, /1967
(m)   to Carcroft Central Workshops 21/2/1962; returned, 25/9/1962
(n)   to Carcroft Central Workshops 30/8/1961; returned, 24/1/1962
(p)   to HC, Leeds, West Yorkshire, 19/10/1965; returned, 9/1966
(q)   ex Goldthorpe Colliery, 28/8/1960
(r)   ex Markham Main Colliery, 19/3/1966
(s)   ex Ashington Central Workshops, Northumberland, 5/1/1976; earlier Thorne Colliery
(t)   ex Ashington Central Workshops, Northumberland, 26/9/1975; earlier Thorne Colliery
(u)   ex CE, Hatton, Derbyshire, /1992; earlier Thurcroft Colliery
(v)   ex Askern Colliery, 26/6/1992

(1) to Doncaster Rural District Council, Cusworth Hall Museum, /1969
(2) written off, c/1973
(3) to J. Quentin, Foxhill, Herefordshire, 24/11/1992
(4) abandoned underground, by 12/1993
(5) to Askern Colliery, 20/10/1971
(6) to RJB Mining (UK) Ltd, with site, 21/3/1994
(7) written off, c/1970
(8) to Askern Colliery, 2/2/1970
(9) to Leeds Industrial Museum, Armley Mills, West Yorkshire, 7/12/1992
(10) to Askern Colliery, 14/1/1971
(11) scrapped, /1976 (by 17/11/1976)
(12) to Askern Colliery, 12/1/1971
(13) to J. Quentin, Foxhill, Herefordshire, 8/12/1992
(14) to Hatfield Colliery, 7/1981
(15) to Hatfield Colliery, c6/1993

ROTHERHAM MAIN COKING PLANT, Canklow N73
ex **John Brown & Co Ltd** SK 424905
NE(C) from 1/1/1947; CPD from 1/1/1963. CLOSED 1963
For site description see ROTHERHAM MAIN COLLIERY.
Gauge : 4ft 8½in

-	0-4-0RE	WSO			(a)	(1)

(a) ex John Brown & Co Ltd, with site, 1/1/1947

(1) sold or scrapped, after 6/1965, by 9/1966

ROTHERHAM MAIN COLLIERY, Canklow N74
ex **John Brown & Co Ltd** SK 424905
NE3 from 1/1/1947. CLOSED 5/1954

The colliery and coking plant were served by sidings on the east side of the BR (ex LMSR) line, 1½ miles south of Rotherham Masborough Station. Also served by the BR (ex LNER) Rotherham Main Colliery Branch, which ran south from their Rotherham – Tinsley line, ½ mile south of Rotherham Central Station (1 mile). Dirt tips south-east of the colliery were reached by an internal branch (½ mile), off which a connection served the ROTHERHAM MAIN DISPOSAL POINT (which see). On closure of the colliery the rail system was retained for the coking plant but we cannot say for sure if ownership of the locomotives passed to the plant.

Gauge : 4ft 8½in

ATLAS No.7	0-4-0ST	OC	HE	311	1883	(a)	(1)
ATLAS No.6 (ATLAS No.10)	0-4-0ST	OC	YE	478	1892	(b)	(2)
No.9 (ATLAS No.14)	0-4-0ST	OC	HL	2454	1900	(c)	(3)
No 15 (ATLAS No.16)	0-4-0ST	OC	HL	2490	1901	(d)	(4)
No.7 FITZWILLIAM	0-4-0ST	OC	HC	916	1910	(e)	(5)
No 2	0-4-0ST	OC	YE	118	1869	(f)	(6)
No.8 (D.M.C.No.2)	0-4-0ST	OC	P	701	1898	(g)	(7)

(a) ex John Brown & Co Ltd, with site, 1/1/1947;
   to Aldwarke Main Colliery c/1947; returned, c/1949 (by 15/4/1949)
(b) ex John Brown & Co Ltd, with site, 1/1/1947;
   to Elsecar Central Workshops c10/1954 (after 8/1954); returned c2/1955 (by 5/1955)

(c)   ex John Brown & Co Ltd, with site, 1/1/1947;
       to Elsecar Central Workshops c1/1960 (possibly 29/2/1960); returned.c4/1960;
       to Handsworth Colliery 10/1/1963; returned 3/4/1963
(d)   ex John Brown & Co Ltd, with site, 1/1/1947
(e)   ex Elsecar Colliery, 10/1949
(f)   ex New Stubbin Colliery, after 8/1950. by 12/1950
(g)   ex Aldwarke Main Colliery, 6/1954, by 8/1954

(1)   scrapped, after 28/5/1950, by 28/6/1953
(2)   to New Stubbin Colliery, 5/1956
(3)   to Smithywood Coking Plant, 1/1964
(4)   scrapped on site, c10/1965 (after 1/1965)
(5)   to Elsecar Central Workshops, 9/1952
(6)   to New Stubbin Colliery, after 12/1950. by 5/1952
(7)   to Aldwarke Main Colliery, 2/1955

ROTHERHAM MAIN DISPOSAL POINT, Canklow   N75
Ministry of Fuel & Power   SK 427903
Opened by MFP by 5/1946.   CLOSED 6/1948

This site was connected to the ROTHERHAM MAIN COLLIERY rail system by the line to the dirt tip approximately ¼ mile south-east of the colliery. The Disposal Point, also referred to as ORCHARD ROAD DISPOSAL POINT or CANKLOW SCREENS, operated spasmodically and the disused track was still in-situ in 1963. The Disposal Point locomotive was normally stabled at the loco shed at Rotherham Main Colliery.

Gauge : 4ft 8½in

71442		0-6-0ST	IC	HE	3206	1944	(1)	(1)

Note that it is probable that a replacement MFP locomotive was here from 1/1947 until 6/1948.

(a)   ex WD, Longmoor, Hampshire, after 2/1945, by 5/1946

(1)   to Wath Colliery, 1/1947

ROUNDWOOD CLOSED COLLIERY, Rawmarsh   N76
ex **Dalton Main Collieries Ltd**   SK 450957
NE3 from 1/1/1947.   CLOSED by 1960

Sidings were retained in connection with a tipping dock on the Sheffield & South Yorkshire Navigation. They were located south-east of the BR ex LMSR) line, ¾ mile north east of Parkgate & Rawmarsh Station. A short line passed beneath the parallel BR (ex LNER) line to access the tipping dock. An NCB line ran from these sidings east and then south-east to join a BR branch over which running powers were exercised to enable locomotives from Silverwood Colliery to work coal to Roundwood. During the period up to 1954 locomotives were allocated to shunt at Roundwood on two occasions, at which times they were stabled under a road bridge. We believe that traffic ceased in the late 1950s.

Gauge : 4ft 8½in

	D.M.C. No.2	0-4-0ST	OC	P	701	1898	(a)	(1)
No.17	(No.3)	0-4-0ST	OC	HC	751	1906	(b)	(2)

(a)   ex Dalton Main Collieries Ltd, 1/1/1947
(b)   ex Elsecar Colliery, 12/1952

(1)   to Silverwood Colliery, after 27/5/1950, by 26/5/1951
(2)   to Aldwarke Main Colliery, after 8/1953, by 2/1954

SHAFTON CENTRAL WORKSHOPS, Shafton   N77
National Coal Board   SE 401103

Opened by NE4 by 1959; NE6 from 26/4/1964; BNY from 26/3/1967; HQ 6/1967.
Rail traffic ceased 11/1971; CLOSED after 1971

Located and served by a siding on the east side of the BR (ex LMSR) Dearne Valley Railway, ½ mile south of Shafton Junction.

Gauge : 4ft 8½in

-	0-4-0DM		HC	D1094	1959	New	(1)
No.5	0-6-0ST	IC	P	1303	1913	(a)	Scr /1966

(a)   ex Houghton Main Colliery, 7/1964

(1)   to Grimethorpe Colliery, 4/1/1971

SILKSTONE COMMON COLLIERY, Silkstone Common   N78
ex **Yorkshire Coal & Trading Co Ltd**   SE 285038 and 288042
NE5 from 1/1/1947.   Merged with WENTWORTH SILKSTONE COLLIERY 1/1/1961

The colliery comprised two sites, linked by a tramway of (presumed) narrow gauge. Served by standard gauge sidings alongside the BR (ex LNER) line, ½ mile south west of Silkstone Station. Shunting was by main line locomotives.

SILVERWOOD COLLIERY, Thrybergh   N79
ex **Dalton Main Collieries Ltd**   SK 478939
NE3 from 1/1/1947; SYK from 26/3/1967; SYG from 1/4/1990; NG from 1/9/1993
CLOSED 23/12/1994

Sidings ran south east from the BR (ex joint), Thrybergh Junction to Braithwell Junction line, 2 miles south east of Thrybergh Junction, to serve the colliery and DALTON MAIN COKING PLANT located to the south-east. After 1959, running powers over the BR line to Roundwood were exercised to reach ROUNDWOOD STAITH (which see). Underground locomotives used here included an experimental overhead wire electric system for coal haulage which worked from 1959 to 1968 and a trapped rail system used for supplies and manriding.

Gauge: 4ft 8½in

No.3		0-4-0ST	OC	HC	751	1906	(a)	(1)
	D.M.C. No.5	0-6-0T	OC	HL	2980	1912	(a)	(2)
39	(D.M.C. No.6)	0-6-0T	OC	HL	3676	1927	(a)	(3)
No 31	(D.M.C. No.7)	0-6-0T	OC	AB	1717	1921	(b)	(4)
No.23	(No.47) (75092)	0-6-0ST	IC	HC	1753	1943	(a)	(5)
No.47		0-6-0T	OC	AB	1301	1912	(c)	(6)
	D.M.C. No.2	0-4-0ST	OC	P	701	1898	(d)	(7)
	TERRY No.56	0-6-0DM		HC	D1091	1958	New (e)	(8)
	ALEX No.59	0-6-0DM		HC	D1138	1958	New	(8)
No.10		0-4-0ST	OC	HL	2464	1900	(f)	(9)
	LESLIE No.68	0-6-0DH		S	10181	1964	New (g)	(10)
(D2208)		0-6-0DM		VF	D209	1953		
				DC	2483	1953	(h)	(11)
	WALTER No.69 521/11001	0-6-0DH		HE	6230	1963	(J)	(12)

(a)   ex Dalton Main Collieries Ltd, with site, 1/1/1947
(b)   ex Dalton Main Collieries Ltd, with site, 1/1/1947;
         to Elsecar Central Workshops, 20/8/1958; returned, 6/1959

(c)   ex Thomas Hall & Sons Ltd, dealers, Llansamlet, Glamorgan, c/1948 (by 22/10/1949) earlier at Port of London Authority, No.43

(d)   ex Roundwood Staith, after 6/1950, by 5/1951;
to Elsecar Central Workshops, 2/1953; returned, by 8/1953

(e)   to Elsecar Central Workshops, after 4/1965, by 7/1965; returned, /1966 (after 4/1966)

(f)   ex Aldwarke Main Colliery, after 3/1962, by 7/1963;
to Cadeby Colliery 13/8/1963; returned 19/9/1963

(g)   to Treeton Colliery, 16/3/1981; returned by 8/1981

(h)   ex Cadeby Colliery, after 8/1970, by 28/7/1981

(j)   ex Elsecar Colliery, 18/3/1981

(1)   to Elsecar Colliery, by 26/5/1951
(2)   scrapped by Thos.W. Ward Ltd, 6/1950
(3)   to Barnburgh Colliery, after 31/5/1959, by 8/1959
(4)   scrapped, /1966 (after 4/1966, by 8/1967)
(5)   to Elsecar Central Workshops, after 4/1966, by 9/1966
(6)   to Frank Tingle Ltd, Kilnhurst, after 4/1958
(7)   to Aldwarke Main Colliery, 10/6/1954
(8)   scrapped on site by Ernest Northcliffe & Sons Ltd, c7/1982 (after 10/1981, by 9/1982)
(9)   scrapped, /1966 (after 4/1966)
(10) to Manvers Colliery, after 8/1981, by 3/1982
(11) dismantled 6/1976; scrapped after 28/8/1978, by 1/1979
(parts seen in a Worksop, (Nottinghamshire) scrapyard, 5/1979)
(12) scrapped on site, 2/1984

Gauge : 2ft 0in (surface stockyard)

-	4wDM	MR		(a)	s/s c/1962

(a)   origins and identity unknown; here by 7/1961

Gauge : 3ft 0in (surface stockyard)

KATY	4wDM	MR	40S280	1968	(a)	(1)
No.46	0-6-0DMF	HE	5316	1956	(b)	(2)

(a)   ex New Stubbin Colliery, 2/8/1979
(b)   ex underground, c4/1981

(1)   to Cadeby Colliery, 2/12/1981
(2)   to Kilnhurst Colliery, 7/1983

Gauge : 3ft 0in (underground locomotives)

No.15		0-6-0DMF	HE	4032	1949	New	(1)
No.35	390/6018	0-6-0DMF	HE	4862	1956	New	(2)
No.44		0-6-0DMF	HE	5314	1956	New (a)	(3)
No.45		0-6-0DMF	HE	5315	1956	New	(4)
No.46		0-6-0DMF	HE	5316	1956	New	(5)
No.47		0-6-0DMF	HE	5317	1956	New	(4)
-		0-6-0DMF	HE	5213	1957	New	(6)
No.1	124/9294	4wWEF	Bg	3485	1958		
			EE	2387	1958	New	(7)

No.2	124/9295	4wWEF	Bg	3486	1958		
			EE	2388	1958	New	(8)
No.3	124/9290	4wWEF	Bg	3487	1958		
			EE	2389	1958	New	(7)
No.4	124/9297 142/18831	4wWEF	Bg	3488	1958		
			EE	2390	1958	New	(7)
No.5	124/9298	4wWEF	Bg	3489	1958		
			EE	2391	1958	New	(7)
No.6		4wWEF	Bg	3490	1958		
			EE	2392	1958	New	(7)
No.7	142/9299A	4wWEF	Bg	3491	1958		
			EE	2393	1958	New	(7)
No.41	No.52 PR6011						
	390/6008	0-6-0DMF	HE	5214	1957	(b)	(9)
No.58		0-6-0DMF	HE	4054	1952	(c)	(10)
No.1	390/18833	4wDHF	HE	8504	1981	New	(4)
No.2	390/18834	4wDHF	HE	8952	1981	New	(4)
No.3	90/18835	4wDHF	HE	8955	1981	New	(4)
No.5	390/18837	4wDHF	HE	8956	1981	New	(4)
No.4	390/18836	4wDHF	HE	8957	1981	New	(4)
No.10	390/6014	0-6-0DMF	HE	3517	1948	(d)	(4)
No.6	141/1986	4wBEF	CE	B1575A	1978		
		rep	CE	B3575	1989	(e)	(11)
No.8	141/88	4wBEF	CE	B1575F	1978	(f)	(4)
No.10	141/90	4wBEF	CE	B2959B	1982	(f)	(12)
	-	4wBEF	CE	B3563B	1989	(g)	(11)
		4w-4wBEF	CE	B3875	1992	New	(4)

(a)   to Elsecar Central Workshops, c/1962; returned, 8/10/1962
(b)   ex Cadeby Colliery, (date unknown)
(c)   ex Upton Colliery, West Yorkshire, 1/7/1960
(d)   ex Cadeby Colliery, after 22/2/1987, by 2/4/1987
(e)   ex CE, Hatton, Derbyshire (converted from 2ft 6in gauge), /1989
        earlier Manton Colliery, Nottinghamshire
(f)   ex Brodsworth Colliery, /1990
(g)   ex CE, Hatton, Derbyshire (converted from 2ft 0in gauge), /1989
        earlier Markham Main Colliery

(1)   to Cadeby Colliery, 4/1981; stripped for spares and remainder scrapped c/1982
(2)   to Cadeby Colliery, 11/1956
(3)   to Kilnhurst Colliery, 11/2/1983
(4)   sold, scrapped or abandoned underground, after 12/1994
(5)   to surface stockyard system, c4/1981
(6)   written off, by 7/1978
(7)   to GECT, Newton le Willows, Lancashire, /1978; thence to Easington Colliery, Co. Durham
(8)   to GECT, Newton le Willows, Lancashire, /1978;
        dismantled for spares and remains scrapped, /1981
(9)   to Cadeby Colliery, by /1980
(10)   to Cadeby Colliery (for spares), 4/1981
(11)   to Clipstone Equipment Holding Centre, Nottinghamshire, after 12/1994
(12)   to Prince of Wales Colliery, West Yorkshire, after 1/12/1994
        (thus possibly after RJB takeover)

Gauge : 400mm (Roadrailer trapped rail system - underground)

No.1	390/18827	2adDHF	BGB	DRL50/400/403			
				(plated DRL50/400/533)	1976	New	(1)
No.2	390/18828	2adDHF	BGB	DRL50/400/405	1976	New	(1)
No.3	390/18829	2adDHF	BGB	DRL50/400/409	1978	New	(1)
No.4	390/18830	2adDHF	BGB	DRL50/400/411	1978	New	(1)
No.5	390/18831	2adDHF	BGB	DRL50/400/419	1979	New	(1)
No.6	390/18832	2adDHF	BGB	DRL50/400/421	1979	New	(1)

(1)   sold, scrapped or abandoned underground, after 12/1994

SKIERS SPRING COLLIERY, Wentworth (Part of ROCKINGHAM COLLIERY) N80

SK 368992

Opened by NE5 1952; BNY from 26/3/1967; Rail traffic ceased c9/1975.   CLOSED 11/1979

A drift mine driven at the site of SKIERS SPRING (WENTWORTH NORTH) DISPOSAL POINT (which see for location details) handled part of the output of ROCKINGHAM COLLIERY from 1952 to 1975.

Gauge : 4ft 8½in

	SENTINEL No.4	4wVBT	VCG	S	9557	1953	New	(1)
	SENTINEL No.3	4wVBT	VCG	S	9394	1950	(a)	(2)
	VENTURE	0-4-0DM		JF	22287	1938	(b)	(3)
(No.4)		0-4-0ST	OC	P	1627	1924	(c)	(4)
No.3	(CARBON)	0-4-0ST	OC	YE	479	1892	(d)	(5)
	H.C. No.3	0-4-0ST	OC	HC	1891	1961	New	(6)
	H.C. No.4	0-4-0ST	OC	HC	1892	1961	(e)	(7)
TL13		4wDH		TH	142C	1964		
	built on frame of 4wVBT		VCG	S	9570		(f)	(8)

(a)   ex Barnsley Main Colliery, after 6/1952, by 28/6/1953
(b)   ex Birdwell Central Workshops, after 7/1958, by 5/1959
(c)   ex Barrow Colliery, after 7/1958, by 10/1959
(d)   ex Wharncliffe Silkstone Colliery, 11/1959
(e)   ex Wombwell Colliery, 18/11/1969
(f)   ex Smithy Wood Colliery, after 7/1972, by 23/2/1973

(1)   to Birdwell Central Workshops after 8/1960, by 10/1960
(2)   to Barnsley Main Colliery, 6/8/1953
(3)   to Birdwell Central Workshops, c/1960
(4)   scrapped c/1966 (after 4/1966)
(5)   to Wharncliffe Silkstone Colliery, after 9/1961, by 4/1962
(6)   scrapped on site by Ogden Ltd, c4/5/1975
(7)   to Titanic Salvage Co Ltd, Ellastone, Staffordshire. 29/1/1977
(8)   to North Gawber Colliery, 26/4/1977

Opened by MFP by 1944; OE from 1/4/1952.
North site to NE5 c1953; South site CLOSED c10/1954

Operated on behalf of the Opencast Executive by
       (Sir Lindsay Parkinson & Co Ltd – initially ?)
       William Pepper & Co Ltd - latterly

The two sites were served by sidings on opposite both sides of the BR (ex LMSR) line, north-east of Wentworth & Hoyland Common Station. They replaced the Wentworth (Harley) Stocking Ground rail loading point at that station. The North site was located on the site of the closed Lidgett Colliery. In 1953, a surface drift was driven here and it became SKIERS SPRING Colliery. Ministry of Fuel & Power locomotives from elsewhere requiring repair were sometimes brought to Skiers Spring to be worked on by fitters from Wentworth Workshops (which see).

Reference : "Industrial Railway Record" No.54, IRS (re Lidgett Colliery)

Gauge : 4ft 8½in

71499	0-6-0ST	IC	HC	1776	1944	(a)	(1)
71507	0-6-0ST	IC	RSHN	7161	1944	(b)	(2)
75135	0-6-0ST	IC	HE	3185	1944	(c)	(3)
75173	0-6-0ST	IC	WB	2761	1944	(d)	(4)
71497	0-6-0ST	IC	HC	1774	1944	(e)	(5)
71510	0-6-0ST	IC	RSHN	7164	1944	(f)	(6)
71442	0-6-0ST	IC	HE	3206	1945	(g)	(7)
92	0-4-0T	OC	9E		1892	(h)	(8)
75060	0-6-0ST	IC	RSHN	7096	1943	(j)	(9)
75064	0-6-0ST	IC	RSHN	7100	1943	(k)	(10)
71508	0-6-0ST	IC	RSHN	7162	1944	(m)	(11)
8416	0-6-0T	IC	Ghd		1877	(n)	(12)

(a)   ex MFP, loan to Garswood Hall Collieries Co Ltd, Wigan, Lancashire, c/1945 (by 8/1945)
(b)   ex WD, (Bicester Depot, Oxfordshire ?), 8/1945;
       to British Oak Disposal Point, West Yorkshire, c12/1949 (BRMO issued 1/12/1949);
       returned. /1953 (by 28/6/1953)
(c)   ex MFP, loan to Wath Main Colliery Co Ltd, 5/1947
(d)   ex MFP, loan to Rossington Colliery, /1948
(e)   ex North Gawber Colliery, 7/1948
(f)   ex West Tinsley Disposal Point, 11/1949;
       to Upper Portland Disposal Point, Nottinghamshire, 9/1950; returned, 4/1952
(g)   ex British Oak Disposal Point, West Yorkshire,12/1949 (BRMO issued 16/12/1949)
(h)   ex British Oak Disposal Point, West Yorkshire,1/1950 (BRMO issued 3/1/1950);
       to Darton Disposal Point, /1950; returned, /1951
(j)   ex Hafod Disposal Point, Denbighshire, 8/1950
(k)   ex Upper Portland Disposal Point, Nottinghamshire, 20/9/1950
(m)   ex WB, Stafford, 19/5/1952; earlier at Broomhill Disposal Point, Northumberland
(n)   ex WB, Stafford, 5/1952; earlier at Broomhill Disposal Point, Northumberland

(1)   to Peel Hall Disposal Point, Lancashire, c/1948 (by 3/1948)
(2)   to Watnall Disposal Point, Nottinghamshire, after 7/1953, by 19/12/1954
(3)   to Wellwood Disposal Point, Fife, 9/1947
(4)   to British Oak Disposal Point, West Yorkshire, after 1/7/1951, by 8/1952
(5)   to West Tinsley Disposal Point, 11/1949
(6)   to Wath & Elsecar Disposal Point, by 6/1952
(7)   to West Tinsley Disposal Point, by 9/1951
(8)   to Darton Disposal Point, /1951

(9)   to Pemberton Disposal Point, Lancashire, c1/1951 (after 8/1950)
(10) to Wath & Elsecar Disposal Point after 1/7/1951, by 1/6/1952
(11) to Pemberton Disposal Point, Lancashire, 8/1953
(12) to Watnall Disposal Point, Nottinghamshire, 19/12/1954

SMITHYWOOD COKING PLANT, Chapeltown   N82

ex **Thorncliffe Coal Distillation Ltd** (Subsidiary of **Newton Chambers & Co Ltd**)   SE 366948

NE(C) 1/1/1947; CPD from 1/1/1963.   CLOSED 1986

Served by sidings on the east side of the BR (ex LNER) line, ½ mile north of Ecclesfield Station. The plant was (until the 1960s) also linked by aerial ropeways with Smithywood, Rockingham and Thorncliffe Drift Collieries and to the Thorncliffe Ironworks of Newton Chambers & Co Ltd.

Gauge : 4ft 8½in

The pre-1947 practise of short-term loans of a spare loco from Newton Chambers & Co Ltd continued as required into the NCB era, but we do not have full details.

T.C.D.	0-4-0ST	OC	YE	1891	1923	(a)	(1)
DAVID	0-4-0ST	OC	AB	2148	1946	(b)	(2)
No.9	0-4-0ST	OC	HL	2454	1900	(c)	(3)
(S.W.C.P. NO.1) (WD140)	0-6-0ST	IC	HE	3193	1944		
		reb	HE	3887	1964	(d)	(4)
S.W.C.P. No.2 (WD139)	0-6-0ST	IC	HE	3192	1944		
		reb	HE	3888	1964	(e)	(5)
-	0-6-0DH		EEV	D1250	1967	(f)	(6)
-	0-4-0DE		YE	2729	1958	(g)	(7)
-	4wDH		RH	476139	1963	(h)	(8)
137	0-6-0DH		RR	10220	1965	(j)	(9)
-	0-4-0DH		RR	10203	1964	(k)	(10)

(a)   ex Thorncliffe Coal Distillation Ltd, with site, 1/1/1947;
          to Birdwell Central Workshops, c /1954 (by 5/1954); returned, after 5/1954, by 16/4/1955;
          to Elsecar Central Workshops c5/1959 (by 5/1959); returned c3/1960 (by 9/4/1960)
(b)   ex Tarslag Ltd, Stockton, Co.Durham, /1953
(c)   ex Rotherham Main Coking Plant, 1/1964
(d)   ex HE, Leeds, West Yorkshire, 3/1964;
          earlier WD, Bramley, Hampshire;
          to HE, for repairs, by 6/1968; returned, by 11/1968;
          to Allerton Bywater Central Workshops, West Yorkshire, 5/1970;
          returned after 11/1970, by 1/1971
(e)   ex HE, Leeds, West Yorkshire, 30/10/1964 (earlier WD)
(f)   ex Nailstone Coal Preparation Plant, Leicestershire, c10/10/1977
(g)   ex Barnsley District Coking Co Ltd, Barrow, c10/1976 (by 16/10/1976)
(h)   ex Nypro Ltd, Flixborough, Lincolnshire, 8/1982
(j)   ex Imperial Chemical Industries Ltd, Billingham, Cleveland
          (per TH, Kilnhurst, on hire), 17/12/1984
(k)   ex Darfield Colliery, 20/6/1985

(1)   scrapped, c8/1964 (after 18/7/1964, by 1/1965)
(2)   scrapped c8/1964 (after 18/7/1964)
(3)   to Steelbreaking & Dismantling Co Ltd, Chesterfield, Derbyshire, 9/1972; scrapped 10/1972
(4)   to Main Line Steam Trust, Loughborough, Leicestershire, c6/1978 (after 11/1977, by 7/1978)
(5)   to South Yorkshire Railway Preservation Group, Chapeltown, 11/1985
(6)   to C.F.Booth Ltd, Rotherham, 21/11/1986

(7) to South Yorkshire Railway. Preservation Society, Chapeltown, after 4/1983, by 5/1984
(8) to C.F.Booth Ltd, Rotherham, 24/11/1986, and scrapped, 2/3/1987
(9) to TH, Kilnhurst, off hire, 4/4/1985
(10) to C.F.Booth Ltd, Rotherham, 21/11/1986, and scrapped, 12/1986

Gauge : 4ft 8½in (coke oven locomotives)

-	0-4-0WE	Unknown			(a)	(1)
-	0-4-0WE	GB	2543	1955	New	(2)
-	0-4-0WE	RSHN	7804	1954	(b)	(3)

(a) ex Thorncliffe Coal Distillation Ltd, with site, 1/1/1947
(b) ex Lambton Coking Plant, Co. Durham, 1/1986 (in week ending 31/1/1986)

(1) scrapped, c/1959 (by 7/1961)
(2) to Leeds Industrial Museum, Armley Mills, West Yorkshire, 6/11/1986
(3) to Universal Contracting Ltd, Thurcroft, /1986 (by 1/1987);
     thence to British Steel Corporation, Ravenscraig, Lanarkshire, per TH, Kilnhurst

SMITHY WOOD COLLIERY, Chapeltown N83
ex N.C. Thorncliffe Collieries Ltd
SE 363953

NE5 from 1/1/1947; BNY from 26/3/1967. CLOSED 12/1972

Sidings between the BR (ex LMSR & ex LNER) lines, ¾ miles north of Ecclesfield LMSR & LNER Stations. Shunting was by main line locomotive until c1955.

Gauge : 4ft 8½in

	-	0-4-0DM	JF	22558	1939	(a)	(1)
	SENTINEL No.2	4wVBT	VCG S	9401	1950	(b)	(2)
	SENTINEL No.6	4wVBT	VCG S	9616	1957	New (c)	(3)
	SENTINEL No.5	4wVBT	VCG S	9570	1954	(d)	(4)
	SENTINEL No.1	4wVBT	VCG S	9400	1950	(e)	(5)
TL13		4wDH	TH	142C	1964	New	
	built on frame of 4wVBT	VCG S	9570				(6)
TL24	TL40	4wDH	TH	156C	1965	New	
	built on frame of 4wVBT	VCG S	(9616 ?)				(7)

(a) ex Grange Colliery, after 10/1955, by 9/4/1960
(b) ex Wharncliffe Silkstone Colliery, 1/1956
(c) to Barrow Colliery, after 4/1957, by 7/1958; returned 3/1960;
     to Birdwell Central Workshops 13/4/1961; returned after 11/1961, by 4/1962;
     to Birdwell Central Workshops 10/1/1963; returned, 3/5/1963
(d) ex Barrow Colliery, after 4/1957, by 18/2/1958;
     ex Birdwell Central Workshops, c/1960 (by 3/1960), by 3/1961; returned, 7/3/1961
(e) ex Wombwell Colliery, after 3/1964, by 5/1964

(1) to Wharncliffe Silkstone Colliery, after 6/1963, by 10/1963
(2) to Birdwell Central Workshops, after 4/1958, by 10/1959
(3) dismantled, 4/1965; frame believed to have gone to TH, Kilnhurst, /1965;
     remains scrapped after 1/1966
(4) frame and wheels to TH, Kilnhurst, after 27/8/1964, by 10/1964;
     remains scrapped on site, after 11/1964
(5) to Wombwell Colliery, after 7/1964, by 27/8/1964
(6) to Skiers Spring Colliery, c2/1973 (after 7/1972, by 23/2/1973)
(7) to Barrow Colliery, /1965 (by 1/1966)

STOCKSBRIDGE COLLIERY, Stocksbridge   N84
ex **Samuel Fox & Co Ltd**   SK 265990 (approx)
NE5 from 1/1/1947.   CLOSED 4/1948
Served by sidings at the west end of Stocksbridge Works complex. Shunting was by Samuel Fox & Co Ltd locos.

TANKERSLEY DRIFT MINE, Upper Tankersley
– See **WHARNCLIFFE SILKSTONE COLLIERY**

THORNCLIFFE CENTRAL WORKSHOPS, Chapeltown   N85
ex **N.C. Thorncliffe Collieries Ltd**   SK 352975
NE5 from 1/1/47.   CLOSED c/1952 on the opening of BIRDWELL CENTRAL WORKSHOPS.
Located adjacent to THORNCLIFFE DRIFT, within the Thorncliffe Ironworks complex.
Gauge : 4ft6 8½in

HUDSWELL No.1	0-6-0ST	OC	HC	1213	1916	(a)	(1)
AVON No.3	0-6-0ST	OC	AE	1826	1919	(b)	(2)
RAYMOND	0-4-0ST	OC	HC	810	1967	(c)	(3)
-	0-4-0ST	OC	YE	1889	1923	(d)	(4)

(a)   ex Wombwell Colliery, 4/1947
(b)   ex Wombwell Colliery, c/1948
(c)   ex (Monk Bretton Colliery ?), by 20/4/1948
(d)   ex Barnsley Main Colliery, after 8/1950, by 1/7/1951

(1)   to Wombwell Colliery, c/1948
(2)   to Wombwell Colliery, 13/3/1950
(3)   to Barrow Colliery, /1948
(4)   to Barnsley Main Colliery, after 29/6/1952, by 1/7/1953

THORNCLIFFE DRIFT, Chapeltown   N86
ex **N.C. Thorncliffe Collieries Ltd**   SK 354974
NE5 from 1/1/1947. CLOSED 28/3/1956 on the opening of BARLEY HALL COLLIERY (which see)
Located on the south west side of the BR (ex LNER) line, ½ mile north-west of Chapeltown Station. This was within the Thorncliffe Ironworks complex and was worked by the locomotives of Newton Chambers & Co Ltd. An aerial ropeway connected this mine to SMITHYWOOD COKING PLANT (which see).

THORNE COLLIERY, Moorends   N87
ex **Pease & Partners Ltd**   SE 705160
NE2 from 1/1/1947..   Colliery CLOSED 1956
Washery DCR from 26/3/1967.   Washery CLOSED 1968
Mothballed site sold to **RJB Mining (UK) Ltd**, c1994
Sidings ran south-east for 1 mile from the BR (ex LNER) line, 2 miles north-west of Thorne North Station, to the colliery and an NCB brickworks.

Coal production at this colliery ceased in 1956 due to excessive ingress of water and distortion of the shafts. The mine was then placed on a "care and maintenance" basis. Reconstruction commenced c1982. Amalgamated Construction Co Ltd (AMCO), of Barnsley, was contracted to carry out mining work and used its own narrow gauge locomotives. The work was never completed.

The washery and railway sidings remained in use from 1956 until 1968, when rail traffic ceased.

Locomotives were used underground in connection with development and care and maintenance work.

Gauge: 4ft 8½in

In addition to the locomotives listed below, 200HP 0-6-0DM locomotives were hired from British Railways during 1967 – 1968. These locos were changed frequently.

10		0-4-0ST	OC	HL	2559	1903		
		reb	#			1937	(a)	(1)
	TEES BRIDGE No.6	0-4-0ST	OC	HL	2971	1912	(a)	(2)
No.3		0-6-0ST	IC	P	1219	1910	(b)	(3)
	THORNE No.1	0-6-0ST	IC	HE	3714	1951	New (c)	(4)
	THORNE No.2	0-6-0ST	IC	HE	3804	1953	New (d)	(4)
	BULLCROFT No.2	0-6-0ST	OC	YE	1787	1922	(e)	(4)

Plate on loco read "Rebuilt Tees Iron Works, April 1937".

(a) ex Pease & Partners Ltd, with site, 1/1/1947
(b) ex Pease & Partners Ltd, with site, 1/1/1947;
   to Carcroft Central Workshops, after 6/1963, by 3/1964; returned, after 5/1965, by 4/1966
(c) to Hatfield Colliery; after 5/1956, by 4/1957; returned; after 10/1967, by 4/1968
(d) to Carcroft Central Workshops, c5/1963 (after 9/1961, by 7/1963);
   returned c10/1963 (by 3/1964)
(e) ex Carcroft Central Workshops, c6/1963 (after 7/1962, by 6/1963)

(1) to Thos. W.Ward Ltd, Bolton on Dearne. c6/1956 (after 5/1956, by 4/1957)
(2) dismantled, 10/1955; sold or scrapped, after 4/1957, by 4/1960
(3) scrapped, after 6/1968, by 10/1968
(4) scrapped, after 4/1968, by 10/1968

Gauge : 2ft 0in (surface stockyard)

-		4wDM	RH	221590	1943	(a)	(1)
-		4wDM	HE	3550	1949	New	(2)
TM 964		4wDM	HE	3551	1949	New	(2)

(a) ex Pease & Partners Ltd, with site, 1/1/1947

(1) to Hatfield Colliery after 9/1968, by 5/1971
(2) to Markham Main Colliery, after 7/1961, by 3/1964

Gauge: 2ft 0in (underground locomotives)

390/T/M/3022	0-4-0DMF	HE	3510	1947	New	(1)
-	0-4-0DMF	HE	3511	1947	New	(2)
390/T/M/3029	0-4-0DMF	HE	3512	1947	New	(3)
-	4wDMF	RH	249557	1947	(a)	(4)
-	4wDMF	RH	249561	1947	(a)	(5)
-	4wDMF	RH	268857	1948	(b)	(6)
-	4wDMF	RH	249559	1947	(c)	(4)

	390/T/M/3015	4wDMF	RH	268870	1950	New	
	(Incorporated parts of RH 249561)						(1)
	390/T/M/3008	4wDMF	RH	249565	1947	(d)	(1)
	-	0-4-0DM	RH	370543	1954	New	(7)
	-	0-6-0DMF	HC	DM839	1954	New	(8)
T1	39O/T/M/3036	0-6-0DMF	HC	DM840	1954	New	(9)
T2	390/T/M/3043	0-6-0DMF	HC	DM841	1954	New	(10)
	390/T/M/3050	0-6-0DMF	HC	DM928	1955	New	(11)
No.21		0-6-0DMF	HC	DM980	1955	New	(12)
No.22		0-6-0DMF	HC	DM986	1956	New	(12)
	390/T/M/3057	0-6-0DMF	HC	DM1108	1959	New	(13)

(a)   ex Markham Main Colliery, c/1950
(b)   ex Wentworth Colliery, c/1950
(c)   ex Yorkshire Main Colliery, altered from 2ft 2in gauge, c3/1951
(d)   ex Yorkshire Main Colliery, altered from 2ft 2in gauge, 1/1952

(1)   to Amalgamated Construction Co Ltd (see next section), /1980
(2)   to Rossington Colliery, 10/1947
(3)   abandoned underground, /1980
(4)   to Hatfield Colliery, c/1957
(5)   to RH, Lincoln, c/1950; parts incorporated in new loco RH 268870; remains scrapped
(6)   to Askern Colliery (surface stockyard), 22/5/1957
(7)   to Askern Colliery, c/1958
(8)   to Rossington Colliery, by 3/1954
(9)   to Ashington Central Workshops, Northumberland, 20/12/1974; thence Rossington Colliery
(10)   to Ashington Central Workshops, Northumberland, 12/12/1974; thence Rossington Colliery
(11)   to Carcroft Central Workshops, 17/6/1959
(12)   to Hatfield Colliery, 8/1956
(13)   to Ashington Central Workshops, Northumberland, 12/12/1974; thence to Hatfield Colliery

Gauge: 2ft 0in (Amalgamated Construction Co Ltd–underground roadway refurbishment contract)

Several Thorne Colliery locomotives were taken over in 1980 by the company to use on the above contract. In view of their condition, a further locomotive from outside was obtained and this did most of the work.

-	0-4-0DMF	HE	3510	1947	(a)	(1)
-	4wDMF	RH	268870	1950	(a)	(1)
(Incorporated parts of	RH	249561)				(1)
-	4wDMF	RH	249565	1947	(a)	(1)
-	4wDMF	RH	481552	1947	(b)	(1)

(a)   ex colliery stock, /1980
(b)   ex Stairfoot Plant Depot, by 7/1980;
       earlier NCB, Point of Ayr Colliery, Flintshire.

(1)   to Amalgamated Construction Co Ltd, Barugh Plant Depot, c/1984

THURCROFT COLLIERY, Thurcroft   N88
ex United Steel Companies Ltd   SK 498897
NE1 from 1/1/1947; SYK from 26/3/1967; SYG from 1/4/1990.   CLOSED 5/12/91
Sidings ran north from the end of the short BR (LMSR & LNER Joint) Thurcroft Colliery branch which ran west from its Brantcliffe Junction - Braithwell Junction line, 2 miles north of Dinnington Junction.

There was a rail served tip north of the colliery. A Coking & By-Products Plant, which had been shunted by colliery locomotives, is believed to have closed shortly before 1/1/1947. Use of standard gauge locomotives ceased in 1976. The trapped rail underground locomotives were used for manriding and supplies and the battery locomotives for supplies.

Gauge : 4ft 8½in

In addition to the locomotives listed below, one or more 350HP 0-6-0DE locos was hired from British Railways for the seven weeks prior to 8/3/1977.

	ROTHERVALE No.2	0-6-0ST	OC	YE	2241	1929	(a)	(1)
(75132)	(ROTHERVALE No.5)							
	THURCROFT UNIT No.5	0-6-0ST	IC	HE	3182	1944	(b)	(2)
	ROTHERVALE No.	0-6-0ST	OC	YE	2240	1929	(c)	(3)
	MALTBY No.3	0-6-0T	OC	AB	2029	1937	(d)	(4)
(D2334)	No.33	0-6-0DM		RSHD	8193	1961		
				DC	2715	1961	(e)	(5)
D2332	LLOYD	0-6-0DM		RSHD	8191	1961		
				DC	2713	1961	(f)	(6)

(a)   ex United Steel Companies Ltd, with site, 1/1/1947
(b)   ex WD, after loan to Belgian State Railways, /1947 (possibly before vesting day)
(c)   earlier Treeton Colliery; possibly here c4/1950
(d)   ex Maltby Colliery, after 4/1957, by 4/1958
(e)   ex Manvers Colliery, 8/10/1969
(f)   ex Cadeby Colliery, 14/6/1976;
       to Shireoaks Colliery, Nottinghamshire, 29/6/1981; returned, 3/9/1982

(1)   to Manton Colliery, Nottinghamshire, /1948
(2)   scrapped on site, after 9/1969, by 8/1970
(3)   later Maltby Colliery, after c4/1950
(4)   to Maltby Colliery, 6/1959
(5)   to Dinnington Colliery, 21/6/1985
(6)   to Dinnington Colliery, 19/7/1985

Gauge : 2ft 0in (underground locomotives)

	524/22	4wBEF	CE	B1574H	1978	(a)	(1)
No.3	524/13	4wBEF	CE	B1574C	1978	(b)	(2)
	524/21	4wBEF	CE	B1574G	1978	(c)	(2)

(a)   ex Steetley Colliery, Nottinghamshire, 12/3/1981
(b)   ex Dinnington Colliery, 7/1981
(c)   ex Swadlincote Training Centre, Derbyshire, 20/7/1981

(1)   to CE, Hatton, Derbyshire, /1992; thence to Manton Colliery, Nottinghamshire
(2)   to CE, Hatton, Derbyshire, /1992; thence to Rossington Colliery

Gauge : 400mm (Underground Roadrailer captive track system)

No.2	390/32	2adDHF	BGB	DRL50/400/429	1980	New	(1)
No.1	390/31	2adDHF	BGB	DRL50/400/434	1981	New	(1)
No.3	390/20000	2adDHF	BGB	DRL50/400/416	1979	(a)	(1)
	390/20001	2adDHF	BGB	DRL50/400/417	1979	(a)	(1)

(a)   ex Wath Colliery, 9/2/1983
(b)   ex Wath Colliery, 23/1/1987

(1)   sold, scrapped or abandoned underground, /1992

ex United Steel Companies Ltd, Rothervale Collieries Branch   SK 437877

NE1 from 1/1/1947; SYK from 26/3/1967; SYG from 1/4/1990.   CLOSED 14/12/1990

The colliery was served by sidings at the end or the BR (ex LNER) Treeton Colliery branch, which ran north-east for 2 miles from Orgreaves Junction. This BR line was worked by NCB locos from ORGREAVE COLLIERY (which see). A further line ran north-west to the BR (ex LMSR) line north of Treeton Station. and to rail served tips to the north of this line. The use of standard gauge locomotives ceased c1981, being replaced by a surface conveyor to Orgreave Washery. The locomotive here in 1982 was under repair as part of a training scheme.

Gauge : 4ft 8½in

In addition to the locomotives listed below, 350HP 0-6-0DE locos were hired from British Railways for 15 weeks in 1977. These locos were frequently changed.

	ROTHERVALE No.0	0-6-0ST	IC	BP	1830	1879		
			reb	YE		1910	(a)	(1)
	ROTHERVALE No.7	0-6-0ST	OC	YE	1021	1909	(b)	(2)
	ROTHERVALE No.1	0-6-0ST	OC	YE	2240	1929	(c)	(3)
	HUNTSMAN	0-6-0ST	OC	AB	2018	1936	(d)	(4)
DL4		0-6-0DM		HC	D1152	1959	New	(5)
	ROTHERVALE No.9	0-6-0ST	OC	HC	1347	1918	(e)	(6)
	MANTON No.2	4wVBT	VCG	S	9548	1952	(f)	(7)
No.20		0-6-0T	OC	HC	1731	1942	(g)	(8)
No.34		0-6-0DH		YE	2913	1965	(h)	(9)
No.24	(2322)	0-6-0DM		RSHD	8181	1961		
				DC	2703	1961	(j)	(10)
D2607	521/21	0-6-0DM		HE	5656	1960	(k)	(11)
No.25		0-6-0DH		YE	2939	1965	(m)	(12)
	LESLIE No.68	0-6-0DH		S	10181	1964	(n)	(13)
	-	4wVBT	VCG	S	9629	1955	(p)	(14)

(a)   ex United Steel Companies Ltd, with site, 1/1/1947
(b)   ex United Steel Companies Ltd, with site, 1/1/1947;
         to Orgreave Colliery, 3/1960; returned, 17/3/1961
(c)   ex United Steel Companies Ltd, with site, 1/1/1947;
         to YE, Sheffield, for repairs, by 20/4/1948; returned by 18/4/1949
(d)   ex United Steel Companies Ltd, with site, 1/1/1947;
         to Brookhouse Colliery, /1949 (by 18/4/1949);
         ex Handsworth Colliery, 5/1950 (by 28/5/1950)
(e)   ex Orgreave Colliery, 4/1959;
         to Orgreave Colliery, c24/4/1959; returned c8/1959 (by 11/1959)
(f)   ex Manton Colliery, Nottinghamshire, 14/4/1959
(g)   ex Orgreave Colliery, 10/1965
(h)   ex Brookhouse Colliery, c/1969 (after 9/1968, by 6/1972)
(j)   ex Orgreave Colliery, c/1972 (after 6/1969, by 6/1972);
         to Orgreave Colliery c1/1973 (after 6/1972, by 4/1973); returned, after 4/1973, by 11/1973
(k)   ex Steetley Colliery, Nottinghamshire, 3/12/1975;
         to Steetley Colliery, 12/1975 or 1/1976; returned, /1981
(m)   ex Orgreave Colliery, 7/12/1975
(n)   ex Silverwood Colliery, 16/3/1981
(p)   ex National Railway Museum, York, for repairs, 8/1982

(1)   scrapped by Marple & Gillott Ltd, of Sheffield, 10/1959 (by 5/10/1959)
(2)   to Manton Colliery, Nottinghamshire, 4/1961
(3)   to Maltby Colliery (possibly via Thurcroft Colliery), 4/1950
(4)   to Brookhouse Colliery, either 4/1954 or 5/1954

(5)	to Brookhouse Colliery, 13/12/1969
(6)	to Orgreave Colliery, after 11/1959, by 4/1960
(7)	to Orgreave Colliery, 9/1959
(8)	to Orgreave Colliery, 9/3/1967
(9)	to Manvers Colliery, 14/9/1982
(10)	to Orgreave Colliery after 11/1973, by 4/1974
(11)	to Steetley Colliery, Nottinghamshire, /1981 (4 months after arrival here)
(12)	to Manvers Colliery, 10/3/1982
(13)	to Silverwood Colliery, after 3/1981, by 8/1981
(14)	to TH (further repairs for National Railway Museum), 3/3/1983; thence returned to National Railway Museum, York

Gauge : 1ft 10in (underground locomotives)

EDWIN	524/41	4wBEF	CE	B2966A	1982	New	(1)
CLIFFE	524/50	4wBEF	CE	B2966B	1982	New	(2)

(1)   to CE, Hatton, Derbyshire, /1991; thence to Kiveton Park Colliery
(2)   to CE, Hatton, Derbyshire, /1989; thence to Bentley Colliery

WALESWOOD COKING PLANT, Wales Bar, near Aston   N90
ex **Waleswood Coking Co Ltd** (Subsidiary of **Skinner & Holford Ltd**)   SK 465838
NE(C) from 1/1/1947.   CLOSED 1961 (may have ceased coking 1958)

Sidings south of the BR (ex LNER) line at Waleswood Station served the coking plant and colliery. The last traffic here was from a Briquetting plant operated by Cawood Wharton & Co Ltd. This closed 1/1962 and the track was lifted in that year. We are uncertain if the locomotives remained the property of Area No.1 after the colliery closed but they are listed here as the coking plant wa their main source of work.

Gauge : 4ft 8½in

(WALESWOOD No.1)	0-4-0ST	OC	HC		750	1906		(1)
		reb	HC			1930	(a)	(1)
-	0-4-0ST	OC	HC		829	1908	(a)	(2)
BIRLEY No.5	0-4-0ST	OC	P		1454	1917	(b)	(3)
DAVID	0-4-0DM		JF	22887	1939	(c)		(4)
ROSSINGTON No.1	0-4-0ST	OC	HC		989	1912	(d)	(5)

(a)   ex Skinner & Holford Ltd, with site, 1/1/1947
(b)   ex Handsworth Colliery, 11/1955
(c)   ex Thos.W.Ward Ltd, Templeborough (hire), c4/1957 (after 1/1957)
(d)   ex Dinnington Colliery, after 8/1959, by 8/1960

(1)   to Kiveton Park Colliery, after 14/1/1961, by 12/5/1961
(2)   sold or scrapped, after 1/1957, by 4/1958
(3)   to Shireoaks Colliery, Nottinghamshire, 1/1956
(4)   returned to Thos.W.Ward Ltd, off hire, c/1957 (by 4/1958)
(5)   to Dinnington Colliery, after 10/1961, by 10/1962

WALESWOOD COLLIERY, Wales Bar, near Aston   N91
ex **Skinner & Holford Ltd**   SK 467838
NE1 from 1/1/1947.   CLOSED 5/1948
See WALESWOOD COKING PLANT.

WALESWOOD DISPOSAL POINT, Wales Bar, near Aston N92
Ministry of Fuel & Power SK 466836
Opened by MFP 12/1941. CLOSED c/1949
Operated by **Sir Lindsay Parkinson & Co Ltd**.

This was one of the first two opencast sites created by the Board of Trade Mines Department. A coal disposal point was established adjacent to WALESWOOD COLLIERY. Coal was first loaded to rail on 27/12/1941 and three days later a loco (so far unidentified) was said to be 'on the way'.

Reference : "Sunshine Miners", P.N. Grimshaw, British Coal Opencast, 1992 (pp.58-60)

Gauge : 4ft 8½in

Note that locomotives were used from late 1941 but have not been identified. Three locomotives were seen here c1948, of which two have been identified.

Gauge : 4ft 8½in

39	No.4 JENNIE	0-6-0ST	OC	HC	1636	1929	(a)	(1)
206	ALLENBY	0-6-0ST	IC	MW	1379	1898	(b)	(2)

(a) ex Sir Lindsay Parkinson Ltd, Temple Newsam Plant Depot, Leeds, by 18/4/1949
(b) ex Sir Lindsay Parkinson Ltd, Temple Newsam Plant Depot, Leeds,
       after 24/2/1948, by 8/1948

(1) returned to Sir Lindsay Parkinson Ltd, Temple Newsam Plant Depot, Leeds, by 21/3/1950
(2) returned to Sir Lindsay Parkinson Ltd, Temple Newsam Plant Depot, Leeds, by 8/1950

WARREN VALE DISPOSAL POINT, Rawmarsh N93
Ministry of Fuel & Power approx SK 444967
Opened by MFP 1940s CLOSED 1940s
Operated by **Sir Lindsay Parkinson & Co Ltd**.

Located close to the A633 road, on the site of Warren Vale Colliery (closed after 1938). Rail connection was from the LNER line north of Kilnhurst Colliery, from which a line ran north-west, passing beneath the parallel LMSR line, to reach the site.

Gauge : 4ft 8½in

Details of locomotives used are not known.

WATH COKING PLANT, Wath on Dearne N94
ex **Wath Main Colliery Co Ltd**
NE(C) 1/1/1947; CLOSED 1956
See WATH COLLIERY.

ex **Wath Main Colliery Co Ltd**   SE 439018

NE3 from 1/1/1947; SYK from 26/3/1967.
Merged with MANVERS COLLIERY 1/1/1986; CLOSED 12/1986

The colliery and coking plant were served by sidings on the south side of the BR (ex LMSR) line, ½ mile north-west of Wath North Station, and were also connected to the BR (ex LNER) line west of its Wath Station by a short line which ran south. From c1956, Wath coal was wound at Manvers Colliery but the rail system was retained to dispose of dirt from the Manvers complex at Wath. The connecting line made use of the closed Hull & Barnsley Railway trackbed to link the two sites.

Gauge : 4ft 8½in

No.45	(WATH MAIN COLLIERY No.2)							
		0-6-0T	IC	HE	830	1904	(a)	(1)
No.46	(WATH MAIN COLLIERY No.3)							
		0-6-0ST	OC	AB	1150	1908	(b)	(2)
11	(No.4)	0-6-0ST	OC	YE	1823	1922	(c)	(3)
No.14	(WATH MAIN No.5)	0-6-0ST	OC	YE	2305	1931	(d)	(4)
71442		0-6-0ST	IC	HE	3206	1945	(e)	(5)
3		0-6-0ST	OC	HC	285	1889	(f)	(6)
	-	0-6-0ST	OC	HC	1052	1914	(g)	(7)
No.28		0-6-0ST	OC	HC	1364	1919	(h)	(8)
No.50		4wVBT	VCG	S	9552	1952	(j)	(9)
	DRAKE	0-4-0ST	OC	P	2026	1942	(k)	(10)
39	FREDERICK	0-6-0T	OC	HL	3676	1927	(m)	(11)
No.71		0-6-0DH		HE	6286	1965	New	(12)
D2373	DAWN No.1	0-6-0DM		Sdn		1961	(n)	(13)
D2225	DEBRA	0-6-0DM		VF	D274	1955		
				DC	2548	1955	(p)	(14)

(a)   ex Wath Main Colliery Co Ltd, with site, 1/1/1947;
           to Kilnhurst Colliery after 8/1953, by 9/1954; returned 11/1954
(b)   ex Wath Main Colliery Co Ltd, with site, 1/1/1947;
           to AB, Kilmarnock, Ayrshire, for repairs, c/1947; returned, c/1947
           to Manvers Colliery after 3/1955, by 7/1955; returned after 3/1964, by 5/1964
(c)   ex Wath Main Colliery Co Ltd, with site, 1/1/1947;
           to Manvers Colliery 4/1966; returned 3/1967
           to Manvers Colliery after 9/4/1967, before 8/1967; returned, after 10/1969, by 1/1970
(d)   ex Wath Main Colliery Co Ltd, with site, 1/1/1947
(e)   ex Rotherham Main Disposal Point (loan from MFP), 1/1947
(f)   ex Elsecar Central Workshops, 9/1948
(g)   ex Elsecar Central Workshops, /1950
(h)   ex Appleby-Frodingham Steel Co Ltd, Scunthorpe, Lincolnshire, 6/1950
(j)   ex Barnburgh Colliery, 8/1955;
           to Manvers Colliery, after 4/1957, by 4/1958; returned, after 7/1958, by 5/1959
(k)   ex Eagre Construction Co Ltd, Scunthorpe, Lincolnshire, /1955 (by 10/7/1955)
           believed used on contract for work at Wath and Manvers Collieries.
(m)   ex Manvers Colliery, 5/1963
(n)   ex Manvers Colliery 2/1977 or 3/1977;
           to Manvers Colliery, after 7/1977, by 9/1977; returned, after 5/1978, by 8/1978
(p)   ex Manvers Colliery, 24/6/1974;
           to Manvers Colliery, by 7/1975; returned, 8/12/1976

(1)   to Manvers Colliery, 3/1956
(2)   to Manvers Colliery, 4/1965
(3)   scrapped on site 1/1970

(4)   scrapped after 10/1965, by 4/1966
(5)   to Wath & Elsecar Disposal Point, after 1/1947, by 6/1947
(6)   to New Stubbin Colliery, 5/1949
(7)   to Elsecar Central Workshops, after 1/1953 by 6/1953
(8)   to Barnburgh Colliery, after 1/1953, by 8/1953
(9)   to Manvers Colliery, after 13/5/1961, by 15/7/1961
(10) to Manvers Colliery, 7/1956
(11) to Manvers Colliery, 7/1963
(12) to Manvers Colliery, after 1/8/1974, by 11/1976
(13) to Manvers Colliery, 11/1978
(14) scrapped on site by Wath Skip Hire Ltd, 7/1985

Gauge : 2ft 2in (surface stockyard)

-		4wDM	MR	9695	1952	(a)	(1)
-		4wDM	RH	382808	1955	(b)	(2)

(a)   ex Cadeby Colliery, 11/1959
(b)   ex Kilnhurst Colliery, 16/3/1971

(1)   scrapped, after 9/1985, by 4/1987
(2)   to Barnsley Metropolitan Borough Council, Elsecar Workshops, after 15/2/1989, by 6/5/1990

Gauge : 2ft 2in (underground locomotive-not used here)

	0-4-0DMF	HE	#			
	rep	HE	7223	1971	(a)	(1)

(a)   ex HE, Leeds, West Yorkshire, 24/2/1971;
        earlier at Cadeby Colliery - see Cadeby note # relating to its original identity
(1)   to Fence Central Workshops. /1971 (by 7/1971)

Gauge : 400mm (Roadrailer trapped rail system - underground)

390/20000	2adDHF	BGB	DRL50/400/416	1979	New	(1)
390/20001	2adDHF	BGB	DRL50/400/417	1979	New	(2)

(1)   to Thurcroft Colliery, 9/2/1983
(2)   to Thurcroft Colliery, 23/1/1987

WATH & ELSECAR DISPOSAL POINT, Brampton, near Wath   N96
Ministry of Fuel & Power   SE 417023

Opened by MFP by 1947; OE from 1/4/1952.
Rail Traffic CEASED and site on Care and Maintenance from 8/5/1964;
                         Locomotive stored here 1967-68; CLOSED c1968.
Operated by the following contractors -
        Fleming Adnitt Ltd (until 1950 at least)
        William Pepper & Co Ltd (latterly)
Served by sidings on the north-west side of the BR (ex LNER) Elsecar branch, immediately west of its junction with the Wath-Barnsley line.
Gauge : 4ft 8½in

71442		0-6-0ST	IC	HE	3206	1945	(a)	(1)
	FORTH	0-6-0ST	OC	AB	1844	1924	(b)	(2)

71510		0-6-0ST	IC	RSHN	7164	1944	(c)	(3)	
75064		0-6-0ST	IC	RSHN	7100	1943	(d)	(4)	
	HOYLAND	0-4-0ST	OC	YE	1026	1910	(e)	(5)	
No.1	QUEEN	0-4-0ST	OC	YE	1027	1910	(f)	(6)	
206		0-4-0ST	OC	AB	1601	1918	(g)	(7)	
	R.O.F.16 No.1	0-4-0ST	OC	HC	1727	1941	(h)	(8)	
47445		0-6-0T	IC	HE	1529	1927	(j)	(9)	

(a)   ex Wath Colliery, after 1/1947, by 6/1947
(b)   ex John Mowlem & Co Ltd, Welham Green Plant Depot, Hertfordshire,
       per Dudley Vale Ltd, dealers, London SW1, c/1949 (by 2/1950)
(c)   ex Upper Portland Disposal Point, Nottinghamshire, 4/1952
(d)   ex Skiers Spring Disposal Point, /1952 (by 1/6/1952);
       to West Tinsley Disposal Point, 7/1953 (after 28/6/1953); returned 3/1956
(e)   ex West Tinsley Disposal Point, 7/1956
(f)   ex Darton Disposal Point, /1958 (after 8/1958)
(g)   ex R.S.Hayes Ltd, dealers, Bridgend, Glamorgan, /1958;
       earlier Steel Company of Wales Ltd, Margam Works, Glamorgan
(h)   ex Cox & Danks Ltd, dealers, Bedford. 6/1961;
       earlier Royal Ordnance Factory, Elstow, Bedfordshire
(j)   ex British Oak Disposal Point, West Yorkshire, 1/6/1967

(1)   to British Oak Disposal Point, West Yorkshire, 4/1948
(2)   to Dudley Vale Ltd, off hire,
       c/o Mowlem's Welham Green Plant Depot, Hertfordshire, c/1951 (by 9/6/1951)
(3)   to Bowers Row Disposal Point, West Yorkshire, 30/7/1958
(4)   to West Tinsley Disposal Point, after 3/1956, by 2/1957
(5)   to Darton Disposal Point, /1957
(6)   to Wm. George (Wath) Ltd, Wath, 7/1961
(7)   to Roland Ward & Sons Ltd, Aldham Bridge Works, Wombwell, /1960 (after 7/1960)
(8)   to British Oak Disposal Point, West Yorkshire, 31/10/1964
(9)   to British Oak Disposal Point, West Yorkshire, 6 or 7/5/1968

WENTWORTH DRIFT MINE, Wentworth   N97
National Coal Board   SK 382986

Opened by NE3 c/1949.   CLOSED 11/195

No standard gauge rail connection. Narrow gauge locomotives were used underground and through level drifts to the surface.

Reference : "Industrial Railway Record No.103" (article by A.J. Booth), IRS

Gauge : 2ft 0in (Surface and underground)

-	4wDM	RH	192843	1938	(a)	(1)
-	4wDMF	RH	256275	1948	New	(2)
-	4wDMF	RH	268857	1948	New	(3)
-	4wDMF	RH	249567	1947	(b)	(3)

(a)   ex Westfield Workshops, Rotherham, /1947
(b)   ex Bentley Colliery, altered from 2ft 3in gauge, 11/1949

(1)   to Elsecar Central Workshops, by 20/4/1948 (still there on 2/3/1949, then sold or scrapped)
(2)   to Elsecar Central Workshops, c3/1956
(3)   to Thorne Colliery, /1950
(4)   to Elsecar Central Workshops, 5/1956

WENTWORTH DISPOSAL POINT
(also known as HARLEY STOCKING GROUND), Tankersley   N98
Ministry of Fuel & Power   SK 364988
Opened by MFP 1942.   CLOSED by 1947
Operated by **Sir Lindsay Parkinson & Co Ltd**

A loading point was established in 1942, on the north-west side of the LMSR Wentworth & Hoyland Common Station, to load coal brought in by road from Wentworth Opencast Site (which see). The loading point was the former exchange sidings of the closed Lidget Colliery. By 1947 it had been replaced by SKIERS SPRING (WENTWORTH NORTH & SOUTH) DISPOSAL POINTS a short distance to the north east.

Gauge : 4ft 8½in

	MONTEVIDEO	0-6-0ST	IC	HC	1683	1937	(a)	(1)
55	MONTEVIDEO	0-6-0ST	IC	HC	1683	1937	(a)	(1)
No.196	ANZAC	0-6-0ST	IC	HE	1856	1937	(b)	(2)

(a)   ex Dale Airfield construction (Air Ministry) (1941-1943) contract, Pembrokeshire, 7/1942
(b)   earlier Temple Newsam Opencast operation, Leeds, 6/1943;
      (assumed to have been here from BRMO of 3/1946 "for 196 ANZAC to move from Wentworth
      to Waterloo Colliery, Leeds" (near Temple Newsam Plant Depot))

(1)   later at HC, Leeds, West Yorkshire, for conversion to oil burning, 12/1944;
      thence later to Esna Barrage contract, Egypt
(2)   to Temple Newsam Plant Depot, Leeds, 3/1946

WENTWORTH OPENCAST SITE, Wentworth   N99
Ministry of Fuel & Power   SK 385976??
Opened by MFP, 6/1942.   Rail system CLOSED and removed c1944
Operated by **Sir Lindsay Parkinson & Co Ltd**

A temporary contractors rail system using dumpcars operated within the pit at an early opencast site in Hague Lane, for the removal of overburden. The railway was not connected to the main line and the coal was removed by road to screens and thence to the rail loading point at WENTWORTH DISPOSAL POINT (see above).

Reference : "Sunshine Miners - Opencast Coalmining in Britain 1942-1992, page 51"
            P. Grimshaw, British Coal Opencast, 1992 (p.57)
            "Industrial Railway Record No.54", (article by T.J. Lodge – pp.250-252), IRS

Gauge : 4ft 8½in

Details of locomotives used not known but there is photographic evidence that a later type 13" MW was based here.

WENTWORTH NORTH and SOUTH DISPOSAL POINTS
- see SKIERS SPRING DISPOSAL POINT

WENTWORTH SILKSTONE COLLIERY, Stainborough   N100
ex **Wentworth Silkstone Collieries Ltd**   SE 310034
NE5 from 1/1/1947; BNY from 26/3/1967.   CLOSED 6/1978

A branch ran south west from the BR (ex LNER) line, 1½ miles east of West Silkstone Junction, to the colliery, which was shunted by main line locomotives. It is noteworthy in that the BR era Worsborough banking engines were coaled at the colliery until the line was electrified.

Gauge: 2ft 4in (Temporary miniature line at 1960 Colliery Gala)

-	4-2-2PM	#		c1960	New	Scr c/1960

\#   the loco was built at Wentworth Silkstone Colliery from pit tub parts.

WENTWORTH STORES, Harley   N101
National Coal Board   SK 368980
Opened by OE by 1974.   to **RJB Mining (UK) Ltd**. 30/12/1994

No rail connection. Surplus locomotives were stored off the track in a compound until required for work or disposed of. A workshop here undertook repairs to MFP locomotive parts in the 1940s. The locomotives were dismantled and re-erected at Skiers Spring by Wentworth fitters.

Gauge : 4ft 8½in

-		4wDH		S	10059	1961	(a)	(1)
1/13		0-6-0DH		HE	7410	1976	New (b)	(2)
-		4wDH		Ulk	2005	(pre 1972)	(c)	Scr c/1980
D2182	3/3	0-6-0DM		Sdn		1962	(d)	(3)
D2258		0-6-0DM		RSHD	7879	1957		
				DC	2602	1957	(e)	(4)
6678		0-4-0DH		HE	6678	1968	(f)	(5)
	THE WELSHMAN	0-6-0ST	IC	MW	1207	1890	(g)	(6)
9		0-6-0ST	OC	YE	2521	1952	(h)	(7)

(a)   ex TH, Kilnhurst, 19/12/1974 or 30/12/1974
(b)   to Coalfield Farm Disposal Point, Leicestershire, 30/6/1976;
       ex Swalwell Disposal Point, Co.Durham, in transit, from 30/6/1989 to 10/7/1989;
       to Denby Disposal Point, Derbyshire, 23/3/1991; returned, 9/11/1991;
       to British Oak Disposal Point, West Yorkshire, c11/1992 (by 5/12/1992);
       returned, after 10/5/1993, by 22/5/1993
(c)   ex Denby Disposal Point, Derbyshire, 7/11/1978;
       to Oxcroft Disposal Point, Derbyshire, c/1980; returned, c/1980
(d)   ex Bennerley Disposal Point, Nottinghamshire, 18/3/1982
(e)   ex Bennerley Disposal Point, Nottinghamshire, 17/2/1984
(f)   ex Bennerley Disposal Point, Nottinghamshire, 27/2/1984
(g)   ex Chatterley Whitfield Mining Museum (in liquidation), Staffordshire,
       after 4/1994, by 7/1994
(h)   ex Chatterley Whitfield Mining Museum (in liquidation), Staffordshire,
       after 4/1994, by 23/7/1994

(1)   to Albert (Abram) Disposal Point, Bickershaw, Lancashire, 27/10/1975
(2)   to RJB Mining (UK) Ltd, with site, 30/12/1994
(3)   to Bennerley Disposal Point, Nottinghamshire, 6/5/1983
(4)   to C.F.Booth Ltd, Rotherham, 2/9/1986 and scrapped there, after 8/1/1987, by 22/1/1987
(5)   to Universal Reclamation Ltd (Shropshire Locomotive Collection),
       Coton Farm, Cross Houses, Shropshire, after 22/1/1992, by 24/2/1992

(6)   remained stored here for British Coal Corporation after site passed to RJB Mining (UK) Ltd; then to Yorkshire Mining Museum Trust, Caphouse Colliery, West Yorkshire, by 6/3/1996

(7)   remained stored here for British Coal Corporation after site passed to RJB Mining (UK) Ltd; then to Yorkshire Mining Museum Trust, Caphouse Colliery, West Yorkshire, by 3/7/1995

Gauge : 2ft 6in (stored pending preservation)

-	4wDMF	RH	480679	1961	(a)	(1)

(a)   ex Chatterley Whitfield Mining Museum (in liquidation), Staffordshire, after 14/4/1994, by 28/10/1994

(1)   remained stored here for British Coal Corporation after site passed to RJB Mining (UK) Ltd; then to Yorkshire Mining Museum Trust, Caphouse Colliery, West Yorkshire, by 6/3/1996

Gauge : 2ft 0in (stored pending preservation)

-	4wDMF	RH	441424	1960	(a)	(1)

(a)   ex Chatterley Whitfield Mining Museum (in liquidation), Staffordshire, after 14/4/1994, by 28/10/1994

(1)   remained stored here for British Coal Corporation after site passed to RJB Mining (UK) Ltd; then to Yorkshire Mining Museum Trust, Caphouse Colliery, West Yorkshire, by 6/3/1996

WEST TINSLEY DISPOSAL POINT, Tinsley   N102

Ministry of Fuel & Power   SK 403897

Opened by MFP c/1945; OE from 1/4/1952.   CLOSED 9/1957

Located adjacent to part of the site of, and used the exchange sidings of, the (then recently closed) TINSLEY PARK COLLIERY, which was served by sidings on the south-west side of the BR Sheffield District line, 1½ miles east of Brightside Junction and ½ mile south-east of West Tinsley Station. Site cleared and track removed c12/1957. The site was later incorporated into the Tinsley Park Steelworks of the English Steel Corporation Ltd.

Initial operator not known; site operated by **William Pepper & Co Ltd**, c/1954 - 12/1956 and by **Burnett & Hallamshire (Fuel) Ltd** from 12/1956 until 9/1957.

Gauge : 4ft 8½in

75147		0--0ST	IC	HE	3187	1944	(a)	(1)
71510		0-6-0ST	IC	RSHN	7164	1944	(b)	(2)
71497		0-6-0ST	IC	HC	1774	1944	(c)	(3)
71442		0-6-0ST	IC	HE	3206	1945	(d)	(4)
75064		0-6-0ST	IC	RSHN	7100	1943	(e)	(5)
	HOYLAND	0-4-0ST	OC	YE	1026	1910	(f)	(6)
	ERIC	0-4-0DM		JF	22881	1939	(g)	(7)

(a)   ex Garswood Hall Collieries Ltd, Wigan, Lancashire (where delivered in error), 9/1944
(b)   ex WD, storage at Bicester Depot, Oxfordshire, /1944
(c)   ex Skiers Spring Disposal Point, 11/1949
(d)   ex Skiers Spring Disposal Point, c3/1951
(e)   ex Wath & Elsecar Disposal Point, 7/1953;
        to Wath & Elsecar Disposal Point, 3/1956; returned, after 3/1956, by 2/1957
(f)   ex Newton Chambers Ltd, Chapeltown, 4/1956

(g)   ex Thos W.Ward Ltd, Sheffield, 1/1957 (on hire until purchased in 3/1957);
        earlier Royal Ordnance Factory, Chorley, Lancashire, ROF CHORLEY No.5

(1)   to MFP, Hafod Disposal Point, Denbighshire, 2/1948
(2)   to Skiers Spring Disposal Point 11/1949
(3)   to LMSR Grimethorpe Motive Power Depot, for repairs, 17/5/1950;
        thence to either Alma or Pilsley Disposal Point, Derbyshire, 8/1950
(4)   to Hafod Disposal Point, Denbighshire, 7/1953
(5)   to Maryport Disposal Point, Cumberland, 9/1957
(6)   to Wath & Elsecar Disposal Point, 7/1956
(7)   Burnett & Hallamshire (Fuel) Ltd, Nunnery, Sheffield, 9/1957

WESTFIELD WORKSHOPS, Parkgate   N103
ex **South Yorkshire Mines Drainage Board**   SK 434956 ?

SM&MDB from 1/1/1947 .   CLOSED 9/1957

The Small Mines & Mines Drainage Unit took over management of those South Yorkshire pumping stations that had previously been jointly owned by the colliery companies which benefited from them. It undertook development of WENTWORTH DRIFT (which see) and produced at Westfield a working loco from two that had been purchased.

None of the SM&MDU activities were rail connected but most, if not all, of the pumping stations had narrow gauge track at the shaft tops.

For the record, pumping stations working in 1947 comprised :-
        CAR HOUSE, Rotherham   SK 427941
        ELSECAR, Hemingfield   SK 394009
        ROB ROYD, Dodworth   SK 332044
        STRAFFORD SILKSTONE, Dodworth SK 323041
        WARREN HOUSE, Rawmarsh   SK 444976 (?)
        WESTFIELD, Parkgate   SK 434956
Reference : "Industrial Railway Record No.103", (article by A.J. Booth), IRS
Gauge : 2ft 0in

-	4wDM	RH	192843	1938	(a)	(1)	
-	4wDM	RH	192848	1938	(a)	(2)	

(a)   ex Thos.W.Ward Ltd, Tinsley Depot, Sheffield, /1947:
        originally Baldry Yerburgh & Hutchinson Ltd, contractors

(1)   to Wentworth Drift Mine development, /1947
(2)   parts used in the repair of RH 192843 and remains scrapped, c/1948

WHARNCLIFFE SILKSTONE,COKING PLANT, Tankersley   N104
ex **Wharncliffe Silkstone Colliery Co Ltd**   SK 339997

NE(C) from 1/1/1947.   CLOSED 1957

See WHARNCLIFFE SILKSTONE COLLIERY.

WHARNCLIFFE SILKSTONE,COLLIERY, Tankersley

ex **Wharncliffe Silkstone Colliery Co Ltd**

SK 337998

NE5 from 1/1/1947; BNY from 26/3/1967.   Merged with ROCKINGHAM 6/1967

Served by sidings on the north-west side of the BR (ex LNER) line, ½ mile south west of Birdwell & Hoyland Common Station. Also connected to the end of the BR (ex LMSR) Wharncliffe branch (closed by 1956) at Birdwell & Pilley Wharf. By 1963, locomotive use was spasmodic. At least part of the coking plant was on the south side of the LNER, opposite the colliery. How the extensive sidings there were worked is not known.

TANKERSLEY DRIFT MINE (SK 343988) opened c1947 and was worked as part of this colliery until the early 1950s. It had no rail connection.

Gauge : 4ft 8½in

No.3	(CARBON)		0-4-0ST	OC	YE	479	1892	(a)	(1)
No.4	(OXYGEN)		0-4-0ST	OC	YE	483	1895	(b)	(2)
No.5	(NITROGEN)	TL 21	0-4-0ST	OC	YE	610	1900	(b)	(3)
	RAYMOND		0-4-0ST	OC	HC	810	1907	(c)	(4)
	SENTINEL No.2		4wVBT	VCG	S	9401	1950	New	(5)
	SENTINEL No.4		4wVBT	VCG	S	9557	1953	(d)	(6)
	-		0-4-0DM		JF	22558	1939	(e)	(7)

(a)   ex Wharncliffe Silkstone Colliery Co Ltd, with site, 1/1/1947;
        to Skiers Spring Colliery 11/1959; returned after 9/1961, by 4/1962
(b)   ex Wharncliffe Silkstone Colliery Co Ltd, with site, 1/1/1947
(c)   ex Barrow Colliery, after 9/1949, by 18/1/1953
(d)   ex Birdwell Central Workshops, c12/1960 (by 12/3/1961)
(e)   ex Smithy Wood Colliery, after 6/1963, by 10/1963

(1)   scrapped, after 10/1963, by 7/1964
(2)   scrapped, 1/1953 (after 18/1/1963)
(3)   scrapped, after 10/1963, by 7/1964
(4)   scrapped on site by Thos.W.Ward Ltd, 1/1961
(5)   to Smithy Wood Colliery, 1/1956
(6)   to Birdwell Central Workshops, by 5/1965; dismantled there and
        frame and wheels to TH, Kilnhurst, 3/6/1965, and remains scrapped
(7)   to Barnsley Main Colliery, 4/1966 or 5/1966

WHARNCLIFFE WOODMOOR COKING PLANT, Carlton

ex **Wharncliffe Woodmoor Colliery Co Ltd**

SE

NE(C) from 1/1/1947.

CLOSED 1956

Adjacent to and worked by the locomotives of WHARNCLIFFE WOODMOOR Nos.1-3 COLLIERIES.

WHARNCLIFFE WOODMOOR Nos. 1-3 COLLIERY, Carlton

NE6 from 1/1/1947.

SE 361096

CLOSED 8/1966

N107

The colliery and adjacent coking plant were located at the end of an NCB line which ran for 1½ miles to the colliery from exchange sidings on the BR (ex LMSR) line, south of Royston & Notton Station. A second branch ran east for 1 mile to the site from the BR (ex LNER) line north of Staincross (for Mapplewell) Station.

Gauge : 4ft 8½in

(11478)		0-6-0ST	IC	VF	878	1880		
		reb	Hor			1897	(a)	(1)
WS4	WOODMOOR No.4	0-6-0ST	IC	HC	581	1901		
		reb	YE			1931	(a)	(2)
WS2		0-6-0ST	IC	P&K		1917		
		rep	J.F.Wake	2305	1923		(a)	(3)
	WOOLLEY No.1	0-6-0ST	OC	P	1801	1931	(b)	(4)
	PRINCESS	0-6-0ST	IC	HE	572	1893	(c)	(5)
(WS5)	WATSON	0-6-0ST	OC	HC	1197	1916	(d)	(6)
N1		0-6-0T	IC	HC	1858	1952	New	(7)
WS 1		0-6-0T	OC	HC	1816	1948	(e)	(8)
(WD 159)		0-6-0ST	IC	HE	3209	1940	(f)	(8)

(a)   ex Wharncliffe Woodmoor Colliery Co Ltd, with site, 1/1/1947
(b)   ex Woolley Colliery, West Yorkshire, c1/1948 (by 7/1948);
      to Wharncliffe Woodmoor Nos. 4-5 Colliery, by 9/4/1955; returned, by /1960;
      to P, Bristol, for repairs, c/1960, by 2/1961; returned, after 7/9/1961
(c)   ex Woolley Colliery, West Yorkshire, after 22/6/1949, by 8/1949
(d)   ex Woolley Colliery, West Yorkshire, after 8/1950, by 4/1951
(e)   ex North Gawber Colliery, 8/1954
(f)   ex WD, Bramley Depot, Hampshire, WD 159, 5/1962

(1)   scrapped by Wm. Hardman, 6/1949
(2)   scrapped by Wakefield Metal Traders Ltd, of Wakefield, after 6/1962, by 12/1962
(3)   scrapped, after 6/1968, by 9/1968
(4)   to Wharncliffe Woodmoor Nos. 4-5 Colliery, /1962
(5)   to Woolley Colliery, West Yorkshire, after 9/1949, by 3/1950
(6)   to North Gawber Colliery, 5/1956
(7)   to North Gawber Colliery, 8/1954
(8)   scrapped, after 6/1968, by 9/1968

Gauge : 2ft 0in

-	4wDM	RH	198261	1940	(a)	(1)
18/212	4wDM	RH	223742	1946	(b)	(2)

(a)   ex unknown location, probably here by 5/1956 (originally new to "S.H. Hurst")
(b)   ex unknown location, probably here by 5/1956 (originally new to Ministry of Supply)

(1)   out of use by 12/1965; sold or scrapped, by /1968
(2)   to Woolley Colliery, West Yorkshire, after 5/1956, by 7/1961

WHARNCLIFFE WOODMOOR Nos.4-5 COLLIERY, Carlton N108

earlier known as CARLTON MAIN COLLIERY
ex **Wharncliffe Woodmoor Colliery Co Ltd** SE 376099
NE6 from 1/1/1947; BNY from 26/3/1967. CLOSED 7/1970

The colliery was served by sidings on the west side of the BR (ex LMS) line, 1½ miles south of Royston & Notton Station. Also connected at the south end of the yard to a spur from the BR (ex LNER, former Hull & Barnsley Railway) line.

Gauge : 4ft 8½in

(No.3)		0-6-0ST	OC	HC	434	1895	(a)	(1)
WS3	SALISBURY	0-6-0ST	IC	P	951	1903		
		reb	YE			1928	(a)	(2)
	WOODMOOR No.5	0-6-0T	IC	HC	884	1911		
		rep	YE			1932		
		reb	HC			1953	(b)	(2)
	WOODMOOR No.4	0-6-0ST	OC	P	1801	1931	(c)	(3)
6	(WD104)	0-6-0ST	IC	HE	2886	1943	(d)	(4)
WR28	(WD 158)	0-6-0ST	IC	HE	3208	1945	(e)	(5)
	YORK No.1	0-4-0ST	OC	YE	2474	1949	(f)	(6)

(a) ex Wharncliffe Woodmoor Colliery Co Ltd, with site, 1/1/1947
(b) ex HC, Leeds, West Yorkshire, after 11/1952, by 7/1953:
     earlier North Gawber Colliery
(c) ex Wharncliffe Woodmoor Nos.1-3 Colliery, 9/4/1955;
     to Wharncliffe Woodmoor Nos.1-3 Colliery, /1960; returned, /1962
(d) ex WD, Donnington Depot, Shropshire, WD 104, /1963
(e) ex WD, Chilwell Depot, Nottinghamshire, WD 158, 2/1965
     (per W.H. Arnott Young & Co Ltd, dealers, Bradford)
(f) ex Monk Bretton Colliery, 30/10/1968

(1) scrapped by Wakefield Metal Traders Ltd, 5/1952
(2) scrapped, after 3/1964, by 7/1964
(3) scrapped, 11/1968
(4) to Thos.W.Ward Ltd, Sheffield, for scrap, early 10/1971
(5) to Woolley Colliery, West Yorkshire, 12/2/1971
(6) to South Kirkby Colliery, West Yorkshire, 12/2/1971

WOMBWELL COLLIERY, Wombwell N109

ex **Wombwell Main Co Ltd** SE 384032
NE5 from 1/1/1947; BNY from 26/3/1967. CLOSED 23/5/1969

The colliery and adjacent brickworks were located on the north east side of the BR (ex LMSR) line, north west of Wombwell Station. Also served by an NCB line running southwards from interchange sidings on the BR (ex LNER) line west of Aldham Junction to the colliery. There was an aerial ropeway running west from the colliery to Barrow Coking Plant.

Gauge : 4ft 8½in

An unidentified Sentinel steam locomotive was reputed to have been here in 12/1950. Either S 9400 or S 9401 would seem to be likely contenders to have been here for a short period.

(No.2)	HUDSWELL No.1	0-6-0ST	OC	HC	1213	1916	(a)	(1)
	AVON No.3	0-6-0ST	OC	AE	1826	1919	(b)	(1)
	YORK No.1	0-4-0ST	OC	YE	2474	1949	New	(2)

(SENTINEL No.2)		4wVBT	VCG	S	9401	1950	(c)	(3)	
SENTINEL No.1		4wVBT	VCG	S	9400	1950	(d)	(4)	
H.C.No.4		0-4-0ST	OC	HC	1892	1961	(e)	(5)	
TL3		4wDH		TH	158C	1965	New		
	built on frame of 4wVBT	VCG	S	9557				(6)	

(a) ex Wombwell Main Co Ltd, with site, 1/1/1947;
to Thorncliffe Central Workshops, 4/1947; returned, c/1948
(b) ex Wombwell Main Co Ltd, with site, 1/1/1947;
to Thorncliffe Central Workshops, c/1948; returned, 13/5/1950;
to Dodworth Colliery, after 4/1957, by 7/1958; returned, after 2/1959, by 10/1959;
to Dodworth Colliery, 10/1959 or 11/1959; ex Barrow Colliery, 22/3/1964
(c) ex Birdwell Central Workshops, 10/1959
(d) ex Barnsley Main Colliery, 20/3/1964;
to Smithy Wood Colliery, after 3/1964, by 5/1964; returned, after 7/1964, by 27/8/1964
(e) ex Barnsley Main Colliery, after 5/1966, by 4/1967

(1) scrapped c9/1965 (after 8/1965, by 3/1966)
(2) to Monk Bretton Colliery, after 24/12/1950, by 7/1951
(3) dismantled after 4/1966; frame to TH, 1/7/1966,
remains sold or scrapped, after 4/1967, by 8/1967
(4) dismantled 11/1968, remains scrapped on site, 10/1969
(5) to Skiers Spring Colliery, 18/11/1969
(6) to Dodworth Colliery, after 6/1969, by 2/1970

Gauge : 2ft 1in (surface stockyard)

TL42	4wDM		HE	6273	1965	New	(1)

(1) to/at Houghton Main Colliery, by after 9/1968, by 6/1974
(may have gone via Dearne Valley Colliery)

YORKSHIRE MAIN COLLIERY, Edlington   N110
ex **Doncaster Amalgamated Collieries Ltd**   SK 541991
NE2 from 1/1/1947; DCR from 26/3/1967.   CLOSED 10/1985

Located south of the BR (ex LMSR) Dearne Valley Railway and connected to it east of Edlington Station. Also connected to the end of a short spur from the BR (ex LNER) Aire Junction to Braithwell Junction line north of Warmsworth Goods Station. Locomotives were used underground for manriding, coal and supplies haulage throughout.

Gauge: 4ft 8½in

In addition to the locomotives listed below, 200HP 0-6-0DM locos were hired from British Railways in 1973. These locos were changed frequently.

No.6	CHESTERFIELD	0-6-0ST	IC	MW	1667	1906	(a)	(1)
(No.5	FORWARD)	0-6-0ST	IC	MW	1690	1906	(b)	(2)
No.24	ARTHUR	0-6-0ST	OC	AE	1892	1921	(b)	(3)
(75050)	(No.35) YM/12/51/M	0-6-0ST	IC	RSHN	7086	1943	(b)	(4)
No.18	EDDIE	0-6-0ST	OC	HC	1178	1916	(c)	(5)
46005	YM1248	0-6-0DM		HE	5240	1957	(d)	(6)
31	BEN	0-6-0ST	OC	AE	2069	1935	(e)	(7)
No.34	CLEMENT	0-6-0ST	IC	HE	1983	1940	(f)	(8)
D2611	YM1835	0-6-0DM		HE	5660	1960	(g)	(9)

EDDIE	3219/027	0-4-0DH	HE	7422	1976	New	(10)

(a)   ex Doncaster Amalgamated Collieries Ltd, with site, 1/1/1947;
      to Hickleton Colliery, after /1948, by 9/1949; returned, by 5/1951
(b)   ex Doncaster Amalgamated Collieries Ltd, with site, 1/1/1947
(c)   ex Appleby-Frodingham Steel Co Ltd, Scunthorpe, Lincolnshire, 6/1950
(d)   ex Hatfield Colliery, after 4/1958, by 7/1961
(e)   ex Hickleton Colliery, after 9/1961, by 5/1963
(f)   ex Brodsworth Colliery, after 5/1963, by 7/1963
(g)   ex BR, Goole, East Yorkshire, 5/1968

(1)   to Askern Colliery, after 5/1951, by 2/6/1952
(2)   scrapped, after 5/1959, by 7/1960
(3)   sold or scrapped, after 4/1958, by 5/1959; frame may have been retained as a trolley
(4)   to Askern Colliery, c27/4/1970
(5)   scrapped, /1972 (by 3/1972)
(6)   written off, 7/1/1978; scrapped, after 1/1980, by 8/1982
(7)   sold or scrapped, after 5/1963 by 9/1963
(8)   to Brodsworth Colliery, after 7/1963, by 12/1963
(9)   written off, 3/8/1976; scrapped on site, c12/1976 (after 8/1976, by 4/1978)
(10) to Manvers Colliery, after 9/1985, by 4/2/1986

Gauge : 3ft 0in (surface stockyard)

TR7		0-4-0DMF	HE	3427	1947	(a)	(1)
	390/YM/1803/M	0-6-0DMF	HC	DM715	1951	(b)	(2)

(a)   ex underground, 9/8/1953
(b)   ex underground, 9/8/1953

(1)   scrapped by Ogden Demolition Ltd, /1986 (after 12/7/1986)
(2)   scrapped by Ogden Demolition Ltd, /1986 (after 17/12/1986)

Gauge : 3ft 0in (underground locomotives)

	-	0-4-0DMF	HE	3287	1945	(a)	(1)
	-	0-4-0DMF	HE	3427	1947	New	(2)
YM1660	390/YM/1659/M	0-4-0DMF	HE	3430	1947	New	(3)
	-	4wDMF	RH	249554	1947	New	(4)
	390/YM/1668/M	0-4-0DMF	HE	3614	1948	New (b)	(5)
	390/YM/1675/M	0-4-0DMF	HE	3615	1948	New	(6)
	390/YM/1683/M	0-4-0DMF	HE	3616	1948	New (c)	(7)
	390/YM/1691/M	0-4-0DMF	HE	3617	1948	New	(8)
	-	0-4-0DMF	HE	3618	1948	New	(9)
	390/YM/1763/M	0-6-0DMF	HC	DM710	1950	New	(3)
	390/YM/1771/M	0-6-0DMF	HC	DM711	1950	New	(3)
	390/YM/1789/M	0-6-0DMF	HC	DM712	1951	New	(3)
	390/YM/1787/M	0-6-0DMF	HC	DM713	1951	New	(3)
	390/YM/1795/M	0-6-0DMF	HC	DM714	1951	New	(3)
	390/YM/1803/M	0-6-0DMF	HC	DM715	1951	New	(10)
	390/Y/1707/M	0-6-0DMF	HC	DM776	1953	New	(3)
	390/Y/1715/M	0-6-0DMF	HC	DM777	1953	New	(11)
	390/Y/1723/M	0-6-0DMF	HC	DM778	1953	New	(3)
	390/Y/1731/M	0-6-0DMF	HC	DM779	1953	New	(3)
	390/Y/1739/M	0-6-0DMF	HC	DM780	1953	New	(3)
No.17	390/Y/1747/M	0-6-0DMF	HC	DM781	1953	New (d)	(12)

390/Y/1755/M	0-6-0DMF	HC	DM782	1953	New	(3)
390/Y/1699/M	0-6-0DMF	HC	DM783	1953	New	(3)
390/Y/1811/M	0-6-0DMF	HC	DM860	1954	(e)	s/s
390/YM/2942/M	0-6-0DMF	HC	DM913	1958	(f)	(3)
390/YM/2104/M	0-6-0DMF	HC	DM1008	1957	New	(3)
-	0-6-0DMF	HC	DM1120	1958	New	(13)
390/YM/2973/M	0-6-0DMF	HC	DM1150	1959	New	(3)
-	0-6-0DMF	HE	8523	1977		
		HC	DM1427	1977	New	(3)

(a) ex Doncaster Amalgamated Collieries Ltd, with site, 1/1/1947
(b) to Carcroft Central Workshops, c/1966; returned, c/1967
(c) to Ashington Central Workshops, Northumberland, /1968; returned, 8/1969
(d) to Ashington Central Workshops, Northumberland, 26/5/1977; returned, 7/1978
(e) ex Brodsworth Colliery, 17/10/1955
(f) ex HC, Leeds, West Yorkshire (repairs), 17/7/1958; earlier Brodsworth Colliery

(1) to Brodsworth Colliery, by /1959
(2) to surface, 9/8/1953
(3) abandoned underground, 10/1985
(4) to Brodsworth Colliery, 7/1948
(5) to Bentley Training Centre, 15/7/1979
(6) to Bentley Colliery, c/1986 (by 18/5/1986)
(7) scrapped on site by Ogden Demolition Ltd, /1986
(8) to Bentley Colliery, c/1986
(9) to Brodsworth Colliery, 11/1954
(10) to surface, c/1983 (by 16/4/1983)
(11) dismantled for spares and parts scrapped, /1978; frame cut up, 5/1979
(12) to Ashington Central Workshops, Northumberland, 17/5/1982;
       thence to Ellington Colliery, Northumberland
(13) to Brodsworth Colliery, 11/1958

Gauge : 2ft 2in (Underground locomotives)

	0-4-0DMF	HE	3128	1944	(a)	(1)
-	0-4-0DMF	HE	3488	1946	(a)	(2)
-	4wDMF	RH	249565	1947	New	(3)
-	4wDMF	RH	249559	1947	(b)	(4)

(a) ex Doncaster Amalgamated Collieries Ltd, with site, 1/1/1947
(b) ex Markham Main Colliery, c3/1951 (altered from 2ft 2in gauge ?)

(1) to Hickleton Colliery (surface), 11/1950
(2) to Markham Main Colliery, by /1954
(3) to Thorne Colliery, 1/1952
(4) to Thorne Colliery, c3/1952

INDEX OF LOCATIONS & FORMER OWNERS

LOCOMOTIVE INDEX

NOTES :   Information normally relates to the locomotive as built.

Column 1   Works Number (or original company running number for locomotives built in main line workshops without a works number).

Column 2   Date ex-works where known - this may be a later year than the year of building or the year recorded on the worksplate.

Column 3   Gauge.

Column 4   Wheel arrangement.

	Steam Locomotives :	**Diesel Locomotives :**
Column 5	Cylinder position	Horse power
Column 6	Cylinder size	Engine type #
Column 7	Driving wheel diameter	Weight in working order
Column 8	Either weight in working order and/or Manufacturers type designation	references
Column 9	Page references	

Manufacturers of petrol and diesel engines :

Ailsa	- Ailsa Craig Ltd, Salfords, Redhill, Surrey
Beardmore	- William Beardmore & Co Ltd, Parkhead, Glasgow
Blackstone	- Blackstone & Co Ltd, Rutland Engineering Works, Stamford, Lincs
Caterpillar	- Caterpillar Tractor Co Ltd
Cummins	- Cummins Engine Co Ltd, Shotts, Lanarkshire
Dorman	- W.H.Dorman & Co Ltd, Tixall Rd, Stafford
EE	- English Electric Co Ltd
Ferguson	- Harry Ferguson Ltd, Coventry
Ford	- Ford Motor Co Ltd, Dearborn, Michigan, USA
Fowler	- John Fowler & Co (Leeds) Ltd, Hunslet, Leeds
Gardner	- L. Gardner & Sons Ltd, Barton Hall Engine Works, Patricroft, Manchester
JAP	- J.A.Prestwich Industries Ltd, Northumberland Park, Tottenham, London
Leyland	- Leyland Motors Ltd, Leyland, Lancashire
Lister	- R.A. Lister & Co Ltd, Dursley, Gloucestershire
MAN	- Maschinenfabrik Augsurg-Nürnburg AG, Germany
McLaren	- J. & H. McLaren Ltd, Midland Engineering Works, Leeds
National	- National Gas & Oil Engine Co Ltd,Ashton-under-Lyne, Lancashire
Paxman	- Davey, Paxman & Co Ltd, Colchester,Essex
Perkins	Perkins Engine Co Ltd, Peterborough, Northants
Petters	- Petters Ltd, Staines, Middx
R-R	- Rolls-Royce Ltd, Oil Engine Division, Shrewsbury
Ruston	- Ruston & Hornsby Ltd, Lincoln
Saurer	- Armstrong Saurer Ltd, Newcastle upon Tyne

ANDREW BARCLAY, SONS & CO LTD, CALEDONIA WORKS, AB
Kilmarnock

889	4.7.1901	4ft 8½in	0-4-0ST	OC	12 x 20	3ft2in			39,49,69
1150	13.4.1908	4ft 8½in	0-6-0ST	OC	16 x 24	3ft7in			56,90
1301	21.9.1912	4ft 8½in	0-6-0T	OC	16 x 24	3ft9in			76
1498	23.3.1918	4ft 8½in	0-6-0ST	OC	14 x 22	3ft5in			24,56
1601	23.7.1918	4ft 8½in	0-4-0ST	OC	14 x 22	3ft5in			92
1650	10.12.1919	4ft 8½in	0-4-0ST	OC	14 x 22	3ft5in			49
1654	2.2.1920	4ft 8½in	0-4-0ST	OC	14 x 22	3ft5in			16,39,49,51
1717	7.4.1921	4ft 8½in	0-6-0T	OC	18 x 24	3ft9in			32,34,76
1792	30.11.1923	4ft 8½in	0-6-0ST	OC	14 x 22	3ft7in			19,24,32,47
1844	26.8.1924	4ft 8½in	0-6-0ST	OC	12 x 20	3ft2in			91
2018	23.3.1936	4ft 8½in	0-6-0ST	OC	14 x 22	3ft5in			16,39,70,87
2025	12.11.1936	4ft 8½in	0-6-0ST	OC	15 x 22	3ft5in			31,47,56
2029	19.3.1937	4ft 8½in	0-6-0T	OC	16 x 24	3ft8in			51,86
2148	10.5.1946	4ft 8½in	0-4-0ST	OC	14½x22	3ft5in			81
2195	9.1.1945	4ft 8½in	0-4-0ST	OC	16 x 24	3ft 7in			67
2215	30.1.1947	4ft 8½in	0-6-0ST	IC	18 x 26	4ft3in			51,70
553	19.3.1968	4ft 8½in	0-6-0DH	375hp		Cummins NT400			67

SA DES ATELIERS DE CONSTRUCTION ÉLECTRIQUES
DE CHARLEROI, Charleroi, Belgium

1928	4ft 8½in	0-4-0WE		39

AVONSIDE ENGINE CO LTD, AVONSIDE ENGINE WORKS, Bristol AE

1448	1902	4ft 8½in	0-6-0T	IC	17 x 24	4ft0in			40
1472	1904	4ft 8½in	0-6-0ST	OC	14 x 20	3ft3in			39,70
1792	1918	4ft 8½in	0-4-0ST	OC					45
1819	1919	4ft 8½in	0-6-0ST	OC	14 x 20	3ft3in			24,32
1826	1919	4ft 8½in	0-6-0ST	OC	14 x 20	3ft3in			7,31,83,99
1833	1919	4ft 8½in	0-6-0ST	OC	14½x20				2
1834	1919	4ft 8½in	0-6-0ST	OC	14½x20				2,9
1838	1919	4ft 8½in	0-4-0ST	OC	14 x 20				6
1839	1919	4ft 8½in	0-4-0ST	OC	14 x 20				7
1892	1921	4ft 8½in	0-6-0ST	OC	14½x20	3ft6in			100
1894	1922	4ft 8½in	0-6-0ST	OC	14½x20	3ft3in			72
1895	29.11.1923	4ft 8½in	0-6-0ST	OC	14½x20	3ft6in	B5	33T	39
1920	31.3.1924	4ft 8½in	0-6-0ST	OC	14½x20	3ft6in	B5	33T	29
1948	1924	4ft 8½in	0-6-0ST	OC	14 x 20	3ft6in			43
1949	1924	4ft 8½in	0-6-0ST	OC	14½x20	3ft3in			13
1950	1924	4ft 8½in	0-6-0ST	OC	14½x20				61,72
2069	1935	4ft 8½in	0-6-0ST	OC	12½x16				43,100

ASSOCIATED ELECTRICAL INDUSTRIES LTD AEI

1272	1969	2ft 6in	4wBEF	90hp	DBF12	[Bg 3660]	52

E. E. BAGULEY LTD, Burton on Trent

Bg

3001	1937	4ft 8½in	4wDM	27hp	Lister		
3400	6.8.1953	2ft 0in	4wBEF	90hp	DBF12	[MV 880]	8
3401	26.10.1953	2ft 0in	4wBEF	90hp	DBF12	[MV 881]	8
3485	27.1.1958	3ft 0in	4wWEF	190hp	EM3A2	[EE 2387]	77
3486	6.2.1958	3ft 0in	4wWEF	190hp	EM3A2	[EE 2388]	78
3487	6.3.1958	3ft 0in	4wWEF	190hp	EM3A2	[EE 2389]	78
3488	21.3.1958	3ft 0in	4wWEF	190hp	EM3A2	[EE 2390]	78
3489	2.4.1958	3ft 0in	4wWEF	190hp	EM3A2	[EE 2391]	78
3490	22.4.1958	3ft 0in	4wWEF	190hp	EM3A2	[EE 2392]	78
3491	29.4.1958	3ft 0in	4wWEF	190hp	EM3A2	[EE 2393]	78
3563	29.9.1960	2ft 6in	4wBEF	90hp	DBF12	[MV 1180]	52
3660	1969	2ft 6in	4wBEF	90hp	DBF12	[AEI 1272]	52

BECORIT (MINING) LTD, Ilkeston, Derbyshire

BGB

DRL25/ /	1971	200mm	1adDHF	25hp	Perkins 3.152		26
DRL25/1/202	1970	200mm	1adDHF	25hp	Perkins 3.152		16
DRL25/2/213	1970	200mm	1adDHF	25hp	Perkins 3.152		26
DRL40/1/506	1972	200mm	2adDHF	40hp			26
DRL40/3/513	1973	200mm	2adDHF	40hp			26
DRL40/3/517	1973	200mm	2adDHF	40hp			26
DRL50/200/523	1975	200mm	2adDHF	50hp	Perkins D4.203	5¾T	26
DRL50/200/527	1975	200mm	2adDHF	50hp	Perkins D4.203	5¾T	26
DRL50/200/531	1976	200mm	2adDHF	50hp	Perkins D4.203	5¾T	26
DRL50/200/536	1976	200mm	2adDHF	50hp	Perkins D4.203	5¾T	26
DRL50/400/403	1976	400mm	2adDHF	50hp	Perkins D4.203	5¾T	79
DRL50/400/405	1976	400mm	2adDHF	50hp	Perkins D4.203	5¾T	79
DRL50/400/409	1978	400mm	2adDHF	50hp	Perkins D4.203	5¾T	79
DRL50/400/411	1978	400mm	2adDHF	50hp	Perkins D4.203	5¾T	79
DRL50/400/416	1979	400mm	2adDHF	50hp	Perkins D4.203	5¾T	86,91
DRL50/400/417	1979	400mm	2adDHF	50hp	Perkins D4.203	5¾T	86,91
DRL50/400/419	1979	400mm	2adDHF	50hp	Perkins D4.203	5¾T	79
DRL50/400/421	1979	400mm	2adDHF	50hp	Perkins D4.203	5¾T	79
DRL50/400/429	1980	400mm	2adDHF	50hp	Perkins D4.203	5¾T	86
DRL50/400/434	1981	400mm	2adDHF	50hp	Perkins D4.203	5¾T	86

BLACK, HAWTHORN & CO LTD, Gateshead

BH

1024	c22.9.1890	4ft 8½in	0-6-0ST	OC	14 x 20	3ft7in	18
1115	1896	4ft 8½in	0-6-0ST	IC	12 x 18	3ft0½in	26

BEYER, PEACOCK & CO LTD, Gorton, Manchester

BP

1830	1879	4ft 8½in	0-6-0ST	IC	16 x 22	4ft3in	87
3872	3.12.1896	4ft 8½in	0-6-0ST	IC			18
4392	29.12.1901	4ft 8½in	0-6-0ST	IC	17 x 24	4ft2in	37

BALDWIN LOCOMOTIVE WORKS, Philadelphia, USA

Bwn

46489	1917	4ft 8½in	0-6-0T	OC	16 x 24	4ft0in	65

NEI MINING EQUIPMENT LTD, CLAYTON EQUIPMENT, Hatton   CE

Originally **CLARKE CHAPMAN LTD.**
Note that, to save space, batches of locomotives are indexed as single entries. For example, 5792A-D refers to locomotives 5792A, 5792B, 5792C and 5792D.

B1574A-H	5.1978	2ft 0in	4wBEF	17½hp	4T	CRT 3½	30,42, 47 53,73,86
B1575A-F	3.1978	2ft 6in	4wBEF	17½hp	4T	CRT 3½	15,52,78
B2238A-B	5.1980	2ft 6in	4wBEF	17½hp	4T	CRT 3½	52
B2259	5.1980	2ft 0in	4wBEF	17½hp	4T	CRT 3½	50
B2959B	1.1982	2ft 6in	4wBEF	17½hp	4T	CRT 3½	15,78
B2964A-B	3.1982	2ft 6in	4wBEF	17½hp	4T	CRT 3½	52
B2966A-B	3.1982	1ft10in	4wBEF	17½hp	4T	CRT 3½	10,50,88
B3000A	1983	2ft 6in	4wBEF	17½hp	4T	CRT 3½	38
B3038	3.1983	2ft 3½in	4wBEF	17½hp	4T	CRT 3½	10,30
B3084	1984	2ft 0in	4wBEF	17½hp	4T	CRT 3½	50
B3086	1984	2ft 0in	4wBEF	13hp	3½T		7
B3101A-B	2.1984	2ft 0½in	4wBEF	17½hp	4T	CRT 3½	50,53
B3118C	1983	2ft 6in	4wBEF	Reb. of B3000		38	
B3142B	1984	2ft 6in	4wBEF	17½hp	4T	CRT 3½	52
B3200A-B	1985	2ft 3in	4wBEF	17½hp	5T	CLH5	38
B3206	1985	2ft 1in	4wBEF	17½hp	4T	CRT 3½	38
B3245	1986	2ft 6in	4wBEF	Reb. of B1575E			52
B3249A-B	8.1986	2ft 1in	4wBEF	17½hp	5T	CLH5	42,47
B3269	1986	2ft 6in	4wBEF	17½hp	4T	CRT 3½	3,6
B3271A-B	3.1986	2ft 3½in	4wBEF	17½hp	4T	CRT3½	10
B3309	1986	2ft 6in	4wBEF	Reb. of B3206			38
B3352A	4.1987	3ft 0in	4w-4wBEF	50hp	10T	BoBo	11
B3417	1988	2ft 0in	4wBEF	Reb. of B2259			50
B3434A-B	1988	2ft 6in	4wBEF	17½hp	4T	CEB4	52,53
B3478	1988	3ft 0in	4wBEF	Reb. of B3352A			11
B3484	1988	3ft 0in	4wBEF	Reb. of B3118C			38
B3553	1989	2ft 6in	4wBEF	Reb. of B3309			38
B3563A-B	1989	3ft 0in	4wBEF	17½hp	4T	CRT 3½	62,78
B3568	1989	2ft 3½in	4wBEF	17½hp	4T	CRT 3½	10
B3575	1989	2ft 0in	4wBEF	Reb. of B1575A			78
B3602A-B	1990	2ft 6in	4w-4wBEF	150hp	21T	CB21	42,73
B3603	3.1990	2ft 0in	4w-4wBEF	150hp	21T	CB21	42
B3645A-B	4.1990	2ft 0in	4wBEF	17½hp	4T	CRT 3½	73
B3656	1990	2ft 3½in	4w-4wBEF	50hp	10T	BoBo	10
B3757	1991	2ft 6in	4wBEF	Reb. of B3101B			53
B3773A	1991	2ft 6in	4w-4wBEF	50hp	10T	BoBo	38
B3794B	1991	2ft 0in	4w-4wBEF	50hp	10T	BoBo	62
B3797	1992	2ft 6in	4w-4wBEF	50hp	10T	BoBo	53
B3864A-B	1992	2ft 6in	4wBEF	Reb. of B1574C/G			73
B3875	1992	2ft 6in	4w-4wBEF	50hp	10T	BoBo	78

DREWRY CAR CO LTD, London (see builders indicated)   DC

2483	2.1953	4ft 8 ½in	0-6-0DM	204hp	[VF D209]	19,24,56,76
2484	3.1953	4ft 8½in	0-6-0DM	204hp	[VF D210]	49,56
2529	10.1954	4ft 8½in	0-6-0DM	204hp	[VF D257]	56
2542	8.1955	4ft 8½in	0-6-0DM	204hp	[VF D268]	7
2548	10.1955	4ft 8½in	0-6-0DM	204hp	[VF D274]	4,56,90

2552	11.1955	4ft 8½in	0-6-0DM	204hp	[VF D278]			16,70
2562	5.1956	4ft 8½in	0-6-0DM	204hp	[VF D288]			54,56
2563	5.1956	4ft 8½in	0-6-0DM	204hp	[VF D289]			31
2580	1.1957	4ft 8½in	0-6-0DM	204hp	[RSHD 7867]			51,56
2602	10.1957	4ft 8½in	0-6-0DM	204hp	[RSHD 7879]			92
2620	7.1959	4ft 8½in	0-6-0DM	204hp	[RSHD 7918]			51
2661	3.1960	4ft 8½in	0-6-0DM	204hp	[RSHD 8102]			37,67
2698	2.1961	4ft 8½in	0-6-0DM	204hp	[RSHD 8176]			24,56
2703	3.1961	4ft 8½in	0-6-0DM	204hp	[RSHD 8181]			49,65,70,87
2707	5.1961	4ft 8½in	0-6-0DM	204hp	[RSHD 8185]			56
2708	5.1961	4ft 8½in	0-6-0DM	204hp	[RSHD 8186]			29,32
2709	5.1961	4ft 8½in	0-6-0DM	204hp	[RSHD 8187]			24,29,49
2713	6.1961	4ft 8½in	0-6-0DM	204hp	[RSHD 8191]			19,29,56,86
2715	7.1961	4ft 8½in	0-6-0DM	204hp	[RSHD 8193]			29,51,56,86
2716	8.1961	4ft 8½in	0-6-0DM	204hp	[RSHD 8194]			51,56
2717	8.1961	4ft 8½in	0-6-0DM	204hp	[RSHD 8195]			56
2718	8.1961	4ft 8½in	0-6-0DM	204hp	[RSHD 8196]			4,56

204hp locos had Gardner 8L3 engines; weight in w/o 29¾T.

DONCASTER WORKS, British Railways   Don

[BR D2057]	5.1959	4ft 8½in	0-6-0DM	204hp	Gardner 8L3	30T	37
[BR D2093]	6.1960	4ft 8½in	0-6-0DM	204hp	Gardner 8L3	30T	37

ENGLISH ELECTRIC CO LTD, Preston, Lancashire   EE

871	1932	4ft 8½in	4wWE			54
2387	27.11958	3ft 0in	4wWEF	190hp	EM3A2 [Bg 3485]	77
2388	6.2.1958	3ft 0in	4wWEF	190hp	EM3A2 [Bg 3486]	78
2389	6.3.1958	3ft 0in	4wWEF	190hp	EM3A2 [Bg 3487]	78
2390	21.3.1958	3ft 0in	4wWEF	190hp	EM3A2 [Bg 3488]	78
2391	2.4.1958	3ft 0in	4wWEF	190hp	EM3A2 [Bg 3489]	78
2392	22.4.1958	3ft 0in	4wWEF	190hp	EM3A2 [Bg 3490]	78
2393	29.4.1958	3ft 0in	4wWEF	190hp	EM3A2 [Bg 3491]	78

ENGLISH ELECTRIC CO LTD, VULCAN WORKS, Newton-le-Willows   EEV

D1199	1967	4ft 8½in	0-6-0DH	380hp	Dorman 6QT		56
D1250	1967	4ft 8½in	0-6-0DH	390hp	Dorman 6QT	[EES 3948]	81

F. C. HIBBERD & CO LTD, Park Royal, London   FH

1841	8.2.1934	2ft 0in	4wDM	20hp	National 2D	3½T	11
2224	12.1939	2ft 0in	4wDM	20hp (reconditioned Simplex)			60
2623	12.1942	2ft 0in	4wDM	20hp	National	Simplex 2½T	68

An unidentified loco possibly by FH appears on page 60

FOX, WALKER & CO, ATLAS ENGINE WORKS, Fishponds, Bristol   FW

382	1878	4ft 8½in	0-6-0ST	OC	13 x 20	3ft6in	Type B1	31,34,65

GREENWOOD & BATLEY LTD, ALBION IRONWORKS, Leeds   GB

1348	20.7.1934	4ft 8½in	0-4-0WE	80hp		25,39
2402	26.11.1953	3ft 0in	4wBEF	10hp	3T	59
2543	6.1.1955	4ft 8½in	0-4-0WE	80hp		82
2605	15.12.1955	4ft 8½in	0-4-0WE	80hp		54
2606	15.12.1955	4ft 8½in	0-4-0WE	80hp		54
2838	20.12.1957	2ft1½in	4wBEF	5hp	Type GB2	38
2839	20.12.1957	2ft1½in	4wBEF	5hp	Type GB2	38,47
2842	21.7.1958	2ft1½in	4wBEF	90hp	Type GB14	38
2846	20.12.1957	2ft 2in	4wBEF	5hp	Type GB2	47
2908	13.11.1959	2ft 1in	4wBEF	90hp	Type GB14	38

GATESHEAD WORKS, NER   Ghd

[NER 71]	11.1877	4ft 8½in	0-6-0T	IC	17 x 22	4ft1in	80

HUDSWELL, CLARKE & CO LTD, RAILWAY FOUNDRY, Leeds   HC

285	18.2.1889	4ft 8½in	0-6-0ST	OC	14 x 20	3ft6½in	31,55,65,90
291	22.11.1887	4ft 8½in	0-4-0ST	OC	13 x 18	3ft0½in	25
365	30.7.1890	4ft 8½in	0-6-0ST	IC	16 x 22	3ft10in	13,16
376	31.12.1891	4ft 8½in	0-6-0ST	IC	16 x 22	3ft10in	70
410	23.6.1893	4ft 8½in	0-6-0ST	IC	16 x 22	3ft10in	70
434	9.5.1895	4ft 8½in	0-6-0ST	OC	15 x 20	3ft6in	99
466	26.3.1897	4ft 8½in	0-6-0ST	IC	13 x 20	3ft3in	26
524	3.5.1899	4ft 8½in	0-4-0ST	OC	13 x 18	3ft0in	72
565	28.9.1900	4ft 8½in	0-6-0ST	IC	16 x 22	3ft10in	70
581	3.5.1901	4ft 8½in	0-6-0ST	IC	16 x 22	3ft10in	98
614	18.6.1902	4ft 8½in	0-6-0ST	OC	13 x 18	3ft1in	67
649	14.7.1903	4ft 8½in	0-6-0ST	OC	14 x 20	3ft7in	18
655	18.8.1903	4ft 8½in	0-6-0ST	IC	13 x 20	3ft3½in	72
692	30.5.1904	4ft 8½in	0-6-0ST	OC	14 x 20	3ft7in	18
704	20.12.1904	4ft 8½in	0-6-0ST	OC	14 x 20	3ft7in	18,56,72
750	5.2.1906	4ft 8½in	0-4-0ST	OC	14 x 20	3ft3½in	49,88
751	28.11.1906	4ft 8½in	0-4-0ST	OC	14 x 20	3ft3½in	1,18,32,75,76
810	20.6.1907	4ft 8½in	0-4-0ST	OC	11 x 16	2ft9½in	7,64,83,97
822	17.9.1912	4ft 8½in	0-6-0ST	OC	14 x 20	3ft7in	4,55
829	25.2.1908	4ft 8½in	0-4-0ST	OC	14 x 20	3ft3½in	88
850	11.2.1909	4ft 8½in	0-6-0T	IC	15½x20	3ft4in	9,22
851	2.6.1909	4ft 8½in	0-6-0T	IC	15½x20	3ft4in	9
862	21.9.1909	4ft 8½in	0-4-0ST	OC	14 x 20	3ft3½in	29
884	16.1.1911	4ft 8½in	0-6-0T	IC	15½x20	3ft4in	67,99
905	29.11.1910	4ft 8½in	0-4-0ST	OC	14 x 20	3ft3½in	63
916	29.6.1910	4ft 8½in	0-4-0ST	OC	14 x 20	3ft3½in	1,31,34,74
920	30.9.1910	4ft 8½in	0-6-0ST	IC	18 x 24	4ft1½in	18
925	29.10.1910	4ft 8½in	0-6-0T	IC	15½x20	3ft4in	9
989	17.9.1912	4ft 8½in	0-4-0ST	OC	14 x 20	3ft3½in	22,29,88
1052	8.1.1914	4ft 8½in	0-6-0ST	OC	15 x 22	3ft4in	31,90
1069	6.11.1914	4ft 8½in	0-6-0T	IC	15½x20	3ft4in	67
1077	29.7.1914	4ft 8½in	0-6-0ST	OC	14 x 20	3ft7in	55
1178	21.2.1916	4ft 8½in	0-6-0ST	OC	15 x 22	3ft4in	100
1197	5.6.1916	4ft 8½in	0-6-0ST	OC	14 x 20	3ft7in	67,98
1213	30.10.1916	4ft 8½in	0-6-0ST	OC	15 x 22	3ft7in	83,99

1338	3.6.1918	4ft 8½in	0-4-0ST	OC	14 x 20	3ft3½in	24,32,34,56
1347	30.7.1918	4ft 8½in	0-6-0ST	OC	15 x 22	3ft4in	70,87
1348	9.9.1918	4ft 8½in	0-6-0ST	OC	15 x 22	3ft4in	9
1349	24.10.1918	4ft 8½in	0-6-0ST	OC	15 x 22	3ft4in	40
1364	9.10.1919	4ft 8½in	0-6-0ST	OC	15 x 22	3ft4in	4,90
1368	21.5.1920	4ft 8½in	0-6-0ST	OC	15 x 22	3ft4in	31,65
1523	30.12.1925	4ft 8½in	0-6-0T	OC	16 x 24	3ft9in	34,65
1531	4.9.1924	4ft 8½in	0-6-0T	IC	16 x 24	4ft0in	55
1532	23.7.1924	4ft 8½in	0-6-0ST	OC	15 x 22	3ft7in	17,72
1580	17.1.1927	4ft 8½in	0-6-0ST	IC	15 x 22	3ft4in	56
1629	31.12.1928	4ft 8½in	0-6-0ST	OC	15 x 22	3ft4in	18
1636	24.7.1929	4ft 8½in	0-6-0ST	OC	12 x 18	3ft1½in	89
1683	17.6.1937	4ft 8½in	0-6-0ST	IC	13 x 20	3ft3½in	93
1690	26.11.1937	4ft 8½in	0-6-0T	IC	16 x 24	4ft0in	4,32,56
1727	11.12.1941	4ft 8½in	0-4-0ST	OC	14 x 22	3ft3½in	92
1731	2.11.1942	4ft 8½in	0-6-0T	OC	17 x 24	3ft9in	19,70,87
1753	30.11.1943	4ft 8½in	0-6-0ST	IC	18 x 26	4ft3in	32,76
1774	30.10.1944	4ft 8½in	0-6-0ST	IC	18 x 26	4ft3in	67,80,95
1776	30.11.1944	4ft 8½in	0-6-0ST	IC	18 x 26	4ft3in	26,80
1792	12.1.1946	4ft 8½in	0-6-0ST	IC	18 x 26	4ft3in	32,56
1816	28.10.1948	4ft 8½in	0-6-0T	OC	16 x 24	3ft9in	67,98
1857	25.9.1952	4ft 8½in	0-6-0T	OC	16 x 24	3ft9in	67
1858	30.10.1952	4ft 8½in	0-6-0T	IC	18 x 24	4ft0in	67,98
1889	30.9.1960	4ft 8½in	0-4-0ST	OC	16 x 24	3ft9in	31,64
1890	28.10.1960	4ft 8½in	0-4-0ST	OC	16 x 24	3ft9in	31
1891	18.9.1961	4ft 8½in	0-4-0ST	OC	16 x 24	3ft9in	79
1892	18.9.1961	4ft 8½in	0-4-0ST	OC	16 x 24	3ft9in	6,79,100
1893	25.5.1961	4ft 8½in	0-4-0ST	OC	16 x 24	3ft9in	7
DM630	25.5.1948	2ft 0in	0-6-0DMF	100hp	Gardner 6LW	15T	23,41,60
DM640	4.6.1948	2ft 0in	0-6-0DMF	100hp	Gardner 6LW	15T	41
DM644	5.6.1948	2ft 0in	0-6-0DMF	100hp	Gardner 6LW	15T	41
DM663	21.6.1952	2ft 0in	0-4-0DMF	68hp	Gardner 4LW	10T	50
DM664	31.10.1952	2ft 0in	0-4-0DMF	68hp	Gardner 4LW	10T	50
DM665	2.1.1953	2ft 0in	0-4-0DMF	68hp	Gardner 4LW	10T	50
DM666	30.1.1953	2ft 0in	0-4-0DMF	68hp	Gardner 4LW	10T	50
DM672	22.12.1948	2ft 0in	0-6-0DMF	100hp	Gardner 6LW	15T	41
DM673	4.2.1949	2ft 0in	0-6-0DMF	100hp	Gardner 6LW	15T	41
DM674	16.2.1949	2ft 0in	0-6-0DMF	100hp	Gardner 6LW	15T	41
DM675	4.4.1949	2ft 0in	0-6-0DMF	100hp	Gardner 6LW	15T	23,41
DM676	6.4.1949	2ft 0in	0-6-0DMF	100hp	Gardner 6LW	15T	41
DM677	27.9.1949	2ft 0in	0-6-0DMF	100hp	Gardner 6LW	15T	23,41
DM689	31.8.1948	2ft 0in	0-4-0DMF	68hp	Gardner 4LW	10T	41
DM690	23.6.1948	2ft 0in	0-4-0DMF	68hp	Gardner 4LW	10T	41
DM691	2.7.1948	2ft 0in	0-4-0DMF	68hp	Gardner 4LW	10T	41
DM703	24.11.1947	2ft 1½in	0-4-0DMF	68hp	Gardner 4LW	10T	25,38
DM704	28.2.1948	2ft 1in	0-4-0DMF	68hp	Gardner 4LW	10T	38
DM710	1.8.1950	3ft 0in	0-6-0DMF	100hp	Gardner 6LW	15T	101
DM711	3.8.1950	3ft 0in	0-6-0DMF	100hp	Gardner 6LW	15T	101
DM712	6.6.1951	3ft 0in	0-6-0DMF	100hp	Gardner 6LW	15T	101
DM713	12.6.1951	3ft 0in	0-6-0DMF	100hp	Gardner 6LW	15T	101
DM714	23.7.1951	3ft 0in	0-6-0DMF	100hp	Gardner 6LW	15T	101
DM715	31.7.1951	3ft 0in	0-6-0DMF	100hp	Gardner 6LW	15T	101
DM716	9.2.1951	2ft 0in	0-6-0DMF	100hp	Gardner 6LW	15T	41
DM717	6.3.1951	2ft 0in	0-6-0DMF	100hp	Gardner 6LW	15T	23,41

DM718	6.3.1951	2ft 0in	0-6-0DMF	100hp	Gardner 6LW	15T	3,41
DM721	2.11.1951	3ft 0in	0-6-0DMF	100hp	Gardner 6LW	15T	18
DM722	12.11.1951	3ft 0in	0-6-0DMF	100hp	Gardner 6LW	15T	18
DM723	26.11.1951	3ft 0in	0-6-0DMF	100hp	Gardner 6LW	15T	18
DM741	4.7.1951	2ft 0in	0-4-0DMF	68hp	Gardner 4LW	10T	30,50
DM749	27.7.1949	2ft 0in	0-4-0DMF	68hp	Gardner 4LW	10T	3,23,62
DM750	30.8.1949	2ft 0in	0-4-0DMF	68hp	Gardner 4LW	10T	62
DM751	2.9.1949	2ft 0in	0-4-0DMF	68hp	Gardner 4LW	10T	62
DM752	23.9.1949	2ft 0in	0-4-0DMF	68hp	Gardner 4LW	10T	23,62,73
DM776	5.2.1953	3ft 0in	0-6-0DMF	102hp	Gardner 6LW	15T	101
DM777	8.2.1953	3ft 0in	0-6-0DMF	102hp	Gardner 6LW	15T	101
DM778	31.3.1953	3ft 0in	0-6-0DMF	102hp	Gardner 6LW	15T	101
DM779	14.7.1953	3ft 0in	0-6-0DMF	102hp	Gardner 6LW	15T	101
DM780	2.5.1953	3ft 0in	0-6-0DMF	102hp	Gardner 6LW	15T	101
DM781	4.5.1953	3ft 0in	0-6-0DMF	102hp	Gardner 6LW	15T	101
DM782	12.6.1953	3ft 0in	0-6-0DMF	102hp	Gardner 6LW	15T	102
DM783	19.6.1953	3ft 0in	0-6-0DMF	102hp	Gardner 6LW	15T	102
DM785	3.1.1952	2ft 0in	0-6-0DMF	100hp	Gardner 6LW	15T	41
DM786	29.12.1953	2ft 0in	0-6-0DMF	100hp	Gardner 6LW	15T	41
DM787	1.1.1954	2ft 0in	0-6-0DMF	100hp	Gardner 6LW	15T	42
DM795	26.9.1952	2ft 0in	0-4-0DMF	68hp	Gardner 4LW	10T	62,72
DM796	19.12.1952	2ft 0in	0-6-0DMF	100hp	Gardner 6LW	15T	22,42,62
DM797	16.3.1953	2ft 0in	0-6-0DMF	100hp	Gardner 6LW	15T	42,62
DM798	1.10.1953	2ft 0in	0-6-0DMF	100hp	Gardner 6LW	15T	23,42,62
DM799	28.10.1953	2ft 0in	0-6-0DMF	100hp	Gardner 6LW	15T	42,62
DM800	15.9.1953	2ft 0in	0-6-0DMF	100hp	Gardner 6LW	15T	3,72
DM801	27.2.1954	2ft 0in	0-6-0DMF	100hp	Gardner 6LW	15T	72
DM802	2.3.1954	2ft 0in	0-6-0DMF	100hp	Gardner 6LW	15T	3,72
DM803	2.4.1954	2ft 0in	0-6-0DMF	100hp	Gardner 6LW	15T	72
DM806	23.7.1953	2ft 0in	0-4-0DMF	68hp	Gardner 4LW	10T	50
DM807	31.8.1953	2ft 0in	0-4-0DMF	68hp	Gardner 4LW	10T	50
DM808	24.9.1953	2ft 0in	0-4-0DMF	68hp	Gardner 4LW	10T	50
DM839	26.1.1954	2ft 0in	0-6-0DMF	100hp	Gardner 6LW	15T	72,85
DM840	6.9.1954	2ft 0in	0-6-0DMF	100hp	Gardner 6LW	15T	73,85
DM841	27.9.1954	2ft 0in	0-6-0DMF	100hp	Gardner 6LW	15T	73,85
DM857	26.10.1954	3ft 0in	0-6-0DMF	100hp	Gardner 6LW	15T	15
DM858	28.10.1954	3ft 0in	0-6-0DMF	100hp	Gardner 6LW	15T	15
DM859	1.11.1954	3ft 0in	0-6-0DMF	100hp	Gardner 6LW	15T	15
DM860	3.11.1954	3ft 0in	0-6-0DMF	100hp	Gardner 6LW	15T	15,102
DM889	4.4.1955	2ft 1in	0-6-0DMF	100hp	Gardner 6LW	15T	38
DM903	28.2.1956	3ft 0in	0-6-0DMF	100hp	Gardner 6LW	15T	15
DM904	28.2.1956	3ft 0in	0-6-0DMF	100hp	Gardner 6LW	15T	15,44
DM905	4.4.1956	3ft 0in	0-6-0DMF	100hp	Gardner 6LW	15T	15
DM906	30.3.1956	3ft 0in	0-6-0DMF	100hp	Gardner 6LW	15T	15
DM907	29.5.1956	3ft 0in	0-6-0DMF	100hp	Gardner 6LW	15T	15
DM908	30.5.1956	3ft 0in	0-6-0DMF	100hp	Gardner 6LW	15T	15,44
DM909	27.11.1956	3ft 0in	0-6-0DMF	100hp	Gardner 6LW	15T	15
DM910	25.1.1957	3ft 0in	0-6-0DMF	100hp	Gardner 6LW	15T	15
DM911	30.1.1957	3ft 0in	0-6-0DMF	100hp	Gardner 6LW	15T	15
DM912	28.2.1957	3ft 0in	0-6-0DMF	100hp	Gardner 6LW	15T	15
DM913	30.1.1958	3ft 0in	0-6-0DMF	100hp	Gardner 6LW	15T	15,102
DM914	30.6.1958	3ft 0in	0-6-0DMF	100hp	Gardner 6LW	15T	11,15
DM928	30.6.1955	2ft 0in	0-6-0DMF	100hp	Gardner 6LW	15T	22,62,85
DM929	28.12.1955	2ft 0in	0-6-0DMF	100hp	Gardner 6LW	15T	23,73
DM930	29.12.1955	2ft 0in	0-6-0DMF	100hp	Gardner 6LW	15T	3,73

DM931	20.6.1956	2ft 0in	0-6-0DMF 100hp	Gardner 6LW	15T	3
DM932	28.3.1956	2ft 0in	0-6-0DMF 100hp	Gardner 6LW	15T	42
DM933	31.5.1956	2ft 0in	0-6-0DMF 100hp	Gardner 6LW	15T	23,73
DM934	26.9.1956	2ft 0in	0-6-0DMF 100hp	Gardner 6LW	15T	22,73
DM935	26.9.1956	2ft 0in	0-6-0DMF 100hp	Gardner 6LW	15T	73
DM936	30.10.1956	2ft 0in	0-6-0DMF 100hp	Gardner 6LW	15T	73
DM937	26.11.1956	2ft 0in	0-6-0DMF 100hp	Gardner 6LW	15T	3,73
DM944	30.3.1955	2ft 1in	0-6-0DMF 100hp	Gardner 6LW	15T	38
D955	30.6.1955	4ft 8½in	0-6-0DM 204hp	Gardner 8L3	30T	40
DM956	31.8.1955	2ft 2in	0-6-0DMF 100hp	Gardner 6LW	15T	22,44
DM979	28.3.1956	2ft 0in	0-6-0DMF 100hp	Gardner 6LW	15T	23,44,62
DM980	30.12.1955	2ft 0in	0-6-0DMF 100hp	Gardner 6LW	15T	23,42,85
DM981	1956	2ft 0in	0-6-0DMF 100hp	Gardner 6LW	15T	42
DM982	29.11.1956	2ft 0in	0-6-0DMF 100hp	Gardner 6LW	15T	73
DM983	31.8.1955	2ft 2in	0-6-0DMF 100hp	Gardner 6LW	15T	44
DM986	24.2.1956	2ft 0in	0-6-0DMF 100hp	Gardner 6LW	15T	42,85
DM987	20.12.1956	2ft 0in	0-6-0DMF 100hp	Gardner 6LW	15T	3
DM1008	28.2.1957	3ft 0in	0-6-0DMF 100hp	Gardner 6LW	15T	102
D1068	1.3.1958	4ft 8½in	0-6-0DM 204hp	Gardner 8L3	35½T	14,17
DM1080	30.10.1957	2ft 0in	0-4-0DMF 68hp	Gardner 4LW	10T	42
D1086	1.6.1958	4ft 8½in	0-6-0DM 204hp	Gardner 8L3	35½T	9
D1090	19.3.1958	4ft 8½in	0-6-0DM 204hp	Gardner 8L3	30T	29,65
D1091	29.5.1958	4ft 8½in	0-6-0DM 204hp	Gardner 8L3	30T	32,76
DM1092	20.12.1957	2ft 0in	0-6-0DMF 100hp	Gardner 6LW	15T	62
D1094	27.8.1959	4ft 8½in	0-4-0DM 72hp	Gardner 4LW	13½T	37,76
DM1107	31.12.1959	2ft 0in	0-6-0DMF 102hp	Gardner 6LW	15T	3
DM1108	29.1.1959	2ft 0in	0-6-0DMF 102hp	Gardner 6LW	15T	42
DM1109	23.4.1959	2ft 2in	0-6-0DMF 102hp	Gardner 6LW	15T	44,85
DM1110	11.9.1959	3ft 0in	0-6-0DMF 102hp	Gardner 6LW	15T	15
D1113	29.12.1958	4ft 8½in	0-6-0DM 204hp	Gardner 8L3	35½T	29
D1115	28.11.1958	4ft 8½in	0-6-0DM 204hp	Gardner 8L3	35½T	18
DM1120	24.7.1958	3ft 0in	0-6-0DMF 100hp	Gardner 6LW	15T	11,15,45,102
DM1126	28.7.1958	2ft 0in	0-6-0DMF 100hp	Gardner 6LW	15T	62
D1128	29.10.1958	4ft 8½in	0-6-0DM 204hp	Gardner 8L3	35½T	19,32,34,65
D1138	22.12.1958	4ft 8½in	0-6-0DM 204hp	Gardner 8L3	35½T	76
D1139	13.4.1959	4ft 8½in	0-6-0DM 204hp	Gardner 8L3	35½T	18
DM1140	27.3.1959	2ft 6in	0-6-0DMF 100hp	Gardner 6LW	15T	9
DM1150	27.1.1959	3ft 0in	0-6-0DMF 100hp	Gardner 6LW	15T	102
DM1151	31.8.1959	2ft 0in	0-6-0DMF 100hp	Gardner 6LW	15T	36,73
D1152	25.6.1959	4ft 8½in	0-6-0DM 204hp	Gardner 8L3	35½T	16,87
D1154	29.7.1959	4ft 8½in	0-6-0DM 204hp	Gardner 8L3	35½T	56
DM1167	30.4.1959	2ft 0in	0-4-0DMF 68hp	Gardner 4LW	10T	42
DM1168	28.5.1959	2ft 0in	0-4-0DMF 68hp	Gardner 4LW	10T	42
D1174	1.12.1959	4ft 8½in	0-6-0DM 204hp	Gardner 8L3	35½T	39,49
D1189	25.2.1959	4ft 8½in	0-6-0DM 204hp	Gardner 8L3	35½T	14
D1202	25.8.1961	4ft 8½in	0-6-0DM 204hp	Gardner 8L3	34½T	14
D1204	11.9.1961	4ft 8½in	0-6-0DM 204hp	Gardner 8L3	34½T	18
D1209	6.10.1961	4ft 8½in	0-6-0DM 204hp	Gardner 8L3	34½T	40
D1210	30.10.1961	4ft 8½in	0-6-0DM 204hp	Gardner 8L3	34½T	40
DM1213	28.3.1960	3ft 0in	0-6-0DMF 100hp	Gardner 6LW	15T	15
DM1214	30.3.1960	3ft 0in	0-6-0DMF 100hp	Gardner 6LW	15T	15
DM1215	29.8.1960	3ft 0in	0-6-0DMF 100hp	Gardner 6LW	15T	15
DM1244	29.5.1961	2ft 1in	0-6-0DMF 100hp	Gardner 6LW	15T	38
DM1249	1.8.1961	3ft 0in	0-6-0DMF 100hp	Gardner 6LW	15T	44
DM1250	3.8.1961	3ft 0in	0-6-0DMF 100hp	Gardner 6LW	15T	44

D1259	16.1.1964	4ft 8½in	0-4-0DH	252hp	Cummins NH5-6-1P	34T			37
DM1284	24.3.1962	3ft 0in	0-6-0DMF	102hp	Gardner 6LW	15T			15
DM1285	24.3.1962	3ft 0in	0-6-0DMF	102hp	Gardner 6LW	15T			15
DM1331	30.5.1964	2ft 0in	0-6-0DMF	100hp	Gardner 6LW	15T			3,73
D1340	1.7.1966	4ft 8½in	0-4-0DH	252hp	Cummins NH5-6-1P	34T			43
D1342	13.9.1966	4ft 8½in	0-4-0DH	252hp	Cummins NH5-6-1P	34T			43
DM1365	15.4.1965	2ft 1in	0-6-0DMF	100hp	Gardner 6LW	15T			38
DM1380	31.1.1966	2ft 0in	0-6-0DMF	102hp	Gardner 6LW	15T			3
D1386	5.10.1966	4ft 8½in	0-6-0DH	252hp	Cummins NH5-6-1P	34T			2,40,61,72
DM1393	7.2.1967	2ft 0in	0-6-0DMF	102hp	Gardner 6LW	15T			3,73
DM1395	3.11.1966	2ft 0in	0-6-0DMF	102hp	Gardner 6LW	15T			3
DM1410	1.10.1969	2ft 0in	0-6-0DMF	102hp	Gardner 6LW	15T			42
DM1411	31.10.1969	2ft 0in	0-6-0DMF	102hp	Gardner 6LW	15T			42
DM1425	30.6.1976	2ft 0in	0-6-0DMF	102hp	Gdnr 6LW	15T [HE 7432]			42
DM1427	1977	3ft 0in	0-6-0DMF	102hp	Gdnr 6LW	15T [HE 8523]			102
DM1442	28.3.1980	2ft 0in	0-6-0DMF	102hp	Gdnr 6LW	15T [HE 8842]			73
DM1443	18.6.1980	2ft 0in	0-6-0DMF	102hp	Gdnr 6LW	15T [HE 8843]			73
DM1444	10.10.1980	2ft 0in	0-6-0DMF	102hp	Gdnr 6LW	15T [HE 8844]			42,73

HUNSLET ENGINE CO LTD, Leeds   HE

311	27.2.1883	4ft 8½in	0-4-0ST	OC	13 x 18	3ft1in			1,74
429	4.11.1887	4ft 8½in	0-4-0ST	OC	13 x 18	3ft1in			1
475	23.12.1889	4ft 8½in	0-4-0ST	OC	13 x 18	3ft1in			1
572	13.1.1893	4ft 8½in	0-6-0ST	IC	13 x 18	3ft1in			98
786	24.6.1902	4ft 8½in	0-6-0ST	IC	14 x 18	3ft2½in			39
830	29.4.1904	4ft 8½in	0-6-0T	IC	15½x20	3ft4in			48,56,90
1499	20.9.1926	4ft 8½in	0-6-0ST	IC	14 x 20	3ft4in	29T		30
1529	20.5.1927	4ft 8½in	0-6-0T	IC	18 x 26	4ft 7in	49½T		92
1643	26.10.1929	4ft 8½in	0-6-0ST	OC	14 x 20	3ft4in	30¼T		26,30,67
1724	10.10.1934	4ft 8½in	0-6-0DM	180hp	Paxman-Ricardo	29T			13,17,43
1826	6.2.1939	4ft 8½in	0-6-0ST	IC	16 x 22	3ft9in	38T		17,61
1856	29.10.1937	4ft 8½in	0-6-0ST	IC	12 x 18	3ft 0in	24½T		93
1935	21.12.1938	2ft 0in	4wDM	20hp	Ailsa Craig CF2	3T-6 cwt			15
1983	18.5.1940	4ft 8½in	0-6-0ST	IC		16 x 22	3ft9in	38T	13,43,100
2002	31.7.1939	2ft 3½in	0-4-0DMF	23hp	Gardner 2L2	4½T			10
2008	19.7.1939	2ft 0in	0-4-0DMF	23hp	Gardner 2L2	4½T			72
2081	5.9.1940	4ft 8½in	0-6-0ST	IC	16 x 22	3ft9in	38T		43
2388	1941	4ft 8½in	0-4-0DMF	25hp	Gardner 2L2	4½T			72
2661	26.2.1943	2ft 3½in	0-4-0DMF	50hp	Gardner 4L2	6T-15 cwt			10
2662	15.4.1943	2ft 3½in	0-4-0DMF	50hp	Gardner 4L2	6T-15 cwt			10,22
2663	16.4.1943	2ft 3½in	0-4-0DMF	50hp	Gardner 4L2	6T-15 cwt			10,22
2688	24.2.1943	4ft 8½in	0-6-0ST	IC	16 x 22	3ft9in	38T		61
2704	27.7.1945	4ft 8½in	0-6-0ST	IC	16 x 22	3ft9in	38T		37
2857	31.5.1943	4ft 8½in	0-6-0ST	IC	18 x 28	4ft3in	48T		31
2872	22.9.1943	4ft 8½in	0-6-0ST	IC	18 x 28	4ft3in	48T		9,22
2886	30.11.1943	4ft 8½in	0-6-0ST	IC	18 x 28	4ft3in	48T		99
3128	4.8.1944	2ft 2in	0-4-0DMF	50hp	Gardner 4L2	8T-7 cwt			44,102
3134	3.2.1944	4ft 8½in	0-6-0ST	IC	18 x 26	4ft3in	48T		70
3149	26.2.1945	2ft 0in	0-4-0DMF	50hp	Gardner 4L2	8T-7 cwt			72
3171	12.6.1944	4ft 8½in	0-6-0ST	IC	18 x 28	4ft3in	48T		67
3178	31.7.1944	4ft 8½in	0-6-0ST	IC	18 x 28	4ft3in	48T		37,46
3181	22.8.1944	4ft 8½in	0-6-0ST	IC	18 x 28	4ft3in	48T		7,12
3182	25.8.1944	4ft 8½in	0-6-0ST	IC	18 x 28	4ft3in	48T		86

3185	14.9.1944	4ft 8½in	0-6-0ST	IC	18 x 28	4ft3in	48T		80
3187	1944	4ft 8½in	0-6-0ST	IC	18 x 28	4ft3in	48T		95
3192	1.11.1944	4ft 8½in	0-6-0ST	IC	18 x 28	4ft3in	48T		81
3193	10.11.1944	4ft 8½in	0-6-0ST	IC	18 x 28	4ft3in	48T		81
3206	26.2.1945	4ft 8½in	0-6-0ST	IC	18 x 28	4ft3in	48T		75,80,90
									91,95
3208	14.3.1945	4ft 8½in	0-6-0ST	IC	18 x 28	4ft3in	48T		99
3209	30.31940	4ft 8½in	0-6-0ST	IC	18 x 28	4ft3in	48T		98
3212	30.4.1940	4ft 8½in	0-6-0ST	IC	18 x 28	4ft3in	48T		67
3287	24.12.1946	3ft 0in	0-4-0DMF	50hp		Gardner 4L2			14,44,101
3316	14.1.1946	2ft 2in	0-4-0DMF	50hp		Gardner 4L2			5,60
3344	2.5.1946	2ft 0in	0-4-0DMF	50hp		Gardner 4L2			30
3417	1.7.1947	3ft 0in	0-6-0DMF	100hp		Gardner 6LW			59
3418	5.9.1947	3ft 0in	0-6-0DMF	100hp		Gardner 6LW			59
3426	11.12.1946	3ft 0in	0-4-0DMF	50hp		Gardner 4L2			14
3427	24.7.1947	3ft 0in	0-4-0DMF	50hp		Gardner 4L2			101
3430	27.7.1947	3ft 0in	0-4-0DMF	50hp		Gardner 4L2			101
3431	23.1.1947	3ft 0in	0-6-0DMF	50hp		Gardner 4L2			5,59
3432	17.6.1948	3ft 0in	0-6-0DMF	100hp		Gardner 6LW			20
3433	19.7.1948	3ft 0in	0-6-0DMF	100hp		Gardner 6LW			59
3437	17.1.1946	2ft 3½in	0-4-0DMF	50hp		Gardner 4L2			10
3438	12.3.1946	2ft 3½in	0-4-0DMF	50hp		Gardner 4L2			10
3439	29.3.1946	2ft 3½in	0-4-0DMF	50hp		Gardner 4L2			10
3440	18.4.1946	2ft 3½in	0-4-0DMF	50hp		Gardner 4L2			10
3488	26.7.1946	2ft 2in	0-4-0DMF	50hp		Gardner 4L2			62,72,102
3510	18.3.1947	2ft 0in	0-4-0DMF	50hp		Gardner 4L2			84,85
3511	4.4.1947	2ft 0in	0-4-0DMF	50hp		Gardner 4L2			72,84
3512	3.6.1947	2ft 0in	0-4-0DMF	50hp		Gardner 4L2			84
3514	18.3.1948	3ft 0in	0-6-0DMF	100hp		Gardner 6LW			20
3515	5.3.1948	3ft 0in	0-6-0DMF	100hp		Gardner 6LW			20,59
3516	6.4.1948	3ft 0in	0-6-0DMF	100hp		Gardner 6LW			20,48
3517	16.4.1948	3ft 0in	0-6-0DMF	100hp		Gardner 6LW			20,78
3519	19.3.1947	2ft 0in	0-4-0DMF	50hp		Gardner 4L2			62,72
3520	16.4.1947	2ft 0in	0-4-0DMF	50hp		Gardner 4L2			62,72
3550	4.4.1949	2ft 0in	4wDM	24hp		Hunslet 2HRW			61,84
3551	4.4.1949	2ft 0in	4wDM	24hp		Hunslet 2HRW			62,84
3557	11.6.1948	2ft 0½in	0-4-0DMF	65hp		Gardner 4LW	10T		33,35
3558	11.6.1948	2ft 0½in	0-4-0DMF	65hp		Gardner 4LW	10T		35,66
3573	1.10.1948	2ft 3½in	0-4-0DMF	65hp		Gardner 4LW			10
3574	4.3.1948	2ft 3½in	0-4-0DMF	65hp		Gardner 4LW			10
3575	4.3.1948	2ft 3½in	0-4-0DMF	65hp		Gardner 4LW			10
3576	23.7.1948	2ft 3½in	0-4-0DMF	65hp		Gardner 4LW			10
3577	23.7.1948	2ft 3½in	0-4-0DMF	65hp		Gardner 4LW			10
3578	1.10.1948	2ft 3½in	0-4-0DMF	65hp		Gardner 4LW			10
3594	5.10.1950	4ft 8½in	0-6-0ST	IC	16 x 22	3ft9in	38T		2,72
3608	6.9.1948	2ft 0in	0-4-0DMF	65hp		Gardner 4LW			62
3609	23.7.1948	2ft 0in	0-4-0DMF	65hp		Gardner 4LW			62
3610	6.9.1948	2ft 0in	0-4-0DMF	65hp		Gardner 4LW			62
3611	7.10.1948	3ft 0in	0-4-0DMF	65hp		Gardner 4LW			14
3612	11.10.1948	3ft 0in	0-4-0DMF	65hp		Gardner 4LW			14
3613	19.10.1948	3ft 0in	0-4-0DMF	65hp		Gardner 4LW			14
3614	16.6.1948	3ft 0in	0-4-0DMF	65hp		Gardner 4LW			11,22,101
3615	16.9.1948	3ft 0in	0-4-0DMF	65hp		Gardner 4LW			9,101
3616	24.9.1948	3ft 0in	0-4-0DMF	65hp		Gardner 4LW			101
3617	3.11.1948	3ft 0in	0-4-0DMF	65hp		Gardner 4LW			9,101

3618	17.121948	3ft 0in	0-4-0DMF	65hp		Gardner 4LW		15,101
3685	31.12.1948	4ft 8½in	0-6-0ST	IC	18 x 26	4ft3in	48T	4,56
3701	7.12.1950	4ft 8½in	0-6-0ST	IC	18 x 26	4ft3in	48T	4,55
3713	30.3.1951	4ft 8½in	0-6-0ST	IC	16 x 22	3ft9in	38T	40,43
3714	12.4.1951	4ft 8½in	0-6-0ST	IC	16 x 22	3ft9in	38T	40,84
3782	29.9.1953	4ft 8½in	0-6-0ST	IC	16 x 22	3ft 9in	38T	61
3783	30.9.1953	4ft 8½in	0-6-0ST	IC	16 x 22	3ft 9in	38T	25,46
3788	29.7.1953	4ft 8½in	0-6-0ST	IC	18 x 26	4ft3in	48T	67
3804	9.10.1953	4ft 8½in	0-6-0ST	IC	18 x 26	4ft3in	48T	22,84
3805	27.10.1953	4ft 8½in	0-6-0ST	IC	18 x 26	4ft3in	48T	25
3832	28.7.1955	4ft 8½in	0-6-0ST	IC	18 x 26	4ft3in	48T	4,56
3834	7.11.1955	4ft 8½in	0-6-0ST	IC	18 x 26	4ft3in	48T	56
3856	1.3.1956	4ft 8½in	0-6-0ST	IC	18 x 26	4ft3in	48T	24,32
3884	1963	4ft 8½in	0-6-0ST	IC	18 x 26	4ft3in	48T	18
3887	1964	4ft 8½in	0-6-0ST	IC	18 x 26	4ft3in	48T	81
3888	1964	4ft 8½in	0-6-0ST	IC	18 x 26	4ft3in	48T	81
3889	18.3.1964	4ft 8½in	0-6-0ST	IC	18 x 26	4ft3in	48T	56
3890	27.3.1964	4ft 8½in	0-6-0ST	IC	18 x 26	4ft3in	48T	18
4025	4.4.1949	2ft 0in	0-4-0DMF	65hp		Gardner 4LW	10T	35,53
4026	4.4.1949	2ft 0in	0-4-0DMF	65hp		Gardner 4LW	10T	53
4032	21.12.1947	3ft 0in	0-6-0DMF	100hp		Gardner 6LW	15T-8 cwt	20,77
4035	12.12.1949	3ft 0in	0-6-0DMF	100hp		Gardner 6LW	15T-8 cwt	20
4036	24.6.1953	3ft 0in	0-6-0DMF	100hp		Gardner 6LW	15T-8 cwt	5
4044	14.2.1957	3ft 0in	0-4-0DMF	70hp		Meadows 4DT420 10T		45
4054	23.9.1952	3ft 0in	0-6-0DMF	100hp		Gardner 6LW	15T-8 cwt	20,78
4069	27.7.1950	3ft 0in	0-6-0DMF	100hp		Gardner 6LW	15T-8 cwt	59
4072	30.4.1954	3ft 0in	0-6-0DMF	100hp		Gardner 6LW	15T-8 cwt	5
4073	26.7.1954	3ft 0in	0-6-0DMF	100hp		Gardner 6LW	15T-8 cwt	5
4076	27.11.1950	2ft 1½in	0-4-0DMF	70hp		Meadows 4DT420 10T		47
4077	8.2.1951	2ft 1½in	0-4-0DMF	70hp		Meadows 4DT420 10T		47
4113	28.2.1955	2ft 0in	0-4-0DMF	70hp		Meadows 4DT420 10T		66
4114	30.4.1955	2ft 0in	0-4-0DMF	70hp		Meadows 4DT420 10T		30,66
4115	30.5.1955	2ft 0in	0-4-0DMF	70hp		Meadows 4DT420 10T		60,66
4128	30.8.1949	2ft 0in	0-4-0DMF	70hp		Meadows 4DT420 10T		35
4129	27.9.1950	1ft 11in	0-4-0DMF	70hp		Meadows 4DT420 10T		21,28
4130	27.9.1950	1ft 11in	0-4-0DMF	70hp		Meadows 4DT420 10T		6,21,28
4332	19.12.1950	1ft 11in	0-4-0DMF	65hp		Gardner 4LW	10T	5,21,28,36
4333	19.12.1950	1ft 11in	0-4-0DMF	65hp		Gardner 4LW	10T	21,28
4467	31.3.1953	2ft 0in	4wDMF	21hp		Ailsa Craig 4RFS21		71
4494	29.10.1953	2ft 3½in	0-4-0DMF	65hp		Gardner 4LW	10T	10
4503	30.6.1955	2ft 0in	0-4-0DMF	70hp		Meadows 4DT420 10T		30,66
4513	28.2.1955	4ft 8½in	0-6-0DM	204hp		Gardner 8L3		13
4808	8.10.1950	1ft 11in	0-4-0DMF	65hp		Gardner 4LW	10T	21,28,50
4815	30.4.1955	3ft 0in	0-6-0DMF	100hp		Gardner 6LW	15T-8 cwt	48,59
4816	26.5.1955	3ft 0in	0-6-0DMF	100hp		Gardner 6LW	15T-8 cwt	11,48,59
4817	21.10.1955	3ft 0in	0-6-0DMF	100hp		Gardner 6LW	15T-8 cwt	48,59
4818	13.12.1955	3ft 0in	0-6-0DMF	100hp		Gardner 6LW	15T-8 cwt	48,59
4819	23.2.1956	3ft 0in	0-6-0DMF	100hp		Gardner 6LW	15T-8 cwt	48,59
4862	30.7.1956	3ft 0in	0-6-0DMF	100hp		Gardner 6LW	15T-8 cwt	20,77
4863	13.9.1956	3ft 0in	0-6-0DMF	100hp		Gardner 6LW	15T-8 cwt	20,48
4864	22.11.1956	3ft 0in	0-6-0DMF	100hp		Gardner 6LW	15T-8 cwt	20,48
4865	18.3.1957	3ft 0in	0-6-0DMF	100hp		Gardner 6LW	15T-8 cwt	20
5203	28.2.1957	1ft 11in	0-4-0DMF	65hp		Gardner 4LW	10T	5,21,28,36
5205	19.3.1957	2ft 3½in	0-4-0DMF	65hp		Gardner 4LW	10T	10
5206	29.3.1957	2ft 3½in	0-4-0DMF	65hp		Gardner 4LW	10T	10

5209	31.5.1957	3ft 0in	0-6-0DMF	65hp	Gardner 4LW	10T	20
5210	6.6.1957	3ft 0in	0-6-0DMF	65hp	Gardner 4LW	10T	20
5211	28.6.1957	3ft 0in	0-6-0DMF	65hp	Gardner 4LW	10T	20
5213	28.9.1957	3ft 0in	0-6-0DMF	65hp	Gardner 4LW	10T	77
5214	10.10.1957	3ft 0in	0-6-0DMF	65hp	Gardner 4LW	10T	20,78
5240	14.8.1957	4ft 8½in	0-6-0DM	264hp	National M4AA6	32T	40,100
5314	15.11.1957	3ft 0in	0-6-0DMF	100hp	Gardner 6LW	15T-8cwt	33,48
							59,77
5315	21.11.1957	3ft 0in	0-6-0DMF	100hp	Gardner 6LW	15T-8 cwt	77
5316	24.12.1957	3ft 0in	0-6-0DMF	100hp	Gardner 6LW	15T-8 cwt	48,77
5317	28.1.1958	3ft 0in	0-6-0DMF	100hp	Gardner 6LW	15T-8 cwt	77
5423	27.3.1964	2ft 0in	0-4-0DMF	65hp	Gardner 4LW	10T	35,50
5430	17.2.1958	3ft 0in	0-6-0DMF	100hp	Gardner 6LW	15T-8 cwt	20
5431	18.2.1958	3ft 0in	0-6-0DMF	100hp	Gardner 6LW	15T-8 cwt	20
5514	26.11.1959	2ft 0in	0-4-0DMF	76hp	Meadows 4DT420	11T	30,66
5590	25.3.1964	4ft 8½in	0-6-0DH	260hp	Gardner 8L3B	32T	61
5597	12.5.1961	2ft 0in	0-4-0DMF	66hp	Gardner 4LW	11T	35
5598	15.5.1961	2ft 0in	0-4-0DMF	66hp	Gardner 4LW	11T	35,50
5607	29.7.1960	3ft 0in	0-6-0DMF	100hp	Gardner 6LW	15T-8 cwt	48
5608	31.10.1960	3ft 0in	0-6-0DMF	100hp	Gardner 6LW	15T-8 cwt	59
5647	30.6.1960	4ft 8½in	0-6-0DM	204hp	Gardner 8L3	32T	2,72
5648	8.7.1960	4ft 8½in	0-6-0DM	204hp	Gardner 8L3	32T	2,43
5656	18.10.1960	4ft 8½in	0-6-0DM	204hp	Gardner 8L3	32T	29,35,87
5660	30.1.1961	4ft 8½in	0-6-0DM	204hp	Gardner 8L3	32T	100
5662	21.2.1961	4ft 8½in	0-6-0DM	204hp	Gardner 8L3	32T	9,14
5665	17.3.1961	4ft 8½in	0-6-0DM	204hp	Gardner 8L3	32T	40
6059	1.5.1962	2ft 0in	0-4-0DMF	76hp	Meadows 4DT420	11T	5,35,66
6227	14.11.1963	2ft 2in	0-6-0DMF	100hp	Gardner 6LW	15T-8 cwt	5
6228	4.12.1963	2ft 2in	0-6-0DMF	100hp	Gardner 6LW	15T-8 cwt	5
6230	13.3.1964	4ft 8½in	0-6-0DH	260hp	Gardner 8L3B	32T	34,76
6273	17.5.1965	2ft 1in	4wDM	28hp	Perkins 3152	4T	46,100
6286	22.3.1965	4ft 8½in	0-6-0DH	260hp	Gardner 8L3B	32T	57,90
6287	28.9.1965	4ft 8½in	0-6-0DH	260hp	Gardner 8L3B	32T	24,34
6631	12.1965	1ft 9in	4wDM	28hp	Perkins 3152	4T	31
6643	5.9.1967	4ft 8½in	0-6-0DM	Reb. of ex-BR D2093			37
6645	10.10.1967	4ft 8½in	0-6-0DM	Reb. of ex-BR D2057			37
6654	26.3.1966	3ft 0in	0-6-0DMF	100hp	Gardner 6LW	15½T	48
6661	25.10.1966	4ft 8½in	0-6-0DH	311hp	R-R C8SFL	55T	4,57
6678	26.2.1968	4ft 8½in	0-4-0DH	233hp	R-R C6SFL	35T	94
7099	24.9.1973	3ft 0in	4w-4wDHF	216hp	2 x Gdnr 6LW	21T	45
7222	5.2.1971	2ft 2in	0-4-0DMF	Reb. of HE 4130	65hp	10T	6
7223	24.2.1971	2ft 2in	0-4-0DMF	Reb. of HE 5203	65hp	10T	6,35,91
7274	18.4.1973	2ft 2in	4wDM	29hp	Perkins 3152	3T-12 cwt	26,46
7405	20.9.1974	4ft 8½in	0-4-0DH	252hp	Cummins	37T	61
7410	15.4.1976	4ft 8½in	0-6-0DH	400hp	R-R C8TFL	45T	94
7422	26.3.1976	4ft 8½in	0-4-0DH	252hp	Cummins	37T	57,101
7432	30.6.1976	2ft 0in	0-6-0DMF	100hp	Gardner 6LW	15T [HC DM1425]	42
7480	31.1.1977	2ft 3½in	0-6-0DMF	100hp	Gardner 6LW	15T-8 cwt	10
7482	30.5.1977	2ft 1½in	0-6-0DMF	100hp	Gardner 6LW	15T-8 cwt	3
7483	30.5.1977	2ft 1in	0-6-0DMF	100hp	Gardner 6LW	15T-8 cwt	3
7530	11.11.1977	2ft 2in	4wDM	43hp	Perkins 4203	3T-12 cwt	46
8504	22.3.1981	3ft 0in	4wDHF#	91hp	Perkins 6354	11.4T	78
8505	2.5.1980	3ft 0in	4wDHF#	91hp	Perkins 6354	11.4T	11,52,61
8506	30.6.1980	2ft 6in	4wDHF#	91hp	Perkins 6354	11.4T	52
8507	15.7.1980	2ft 6in	4wDHF#	91hp	Perkins 6354	11.4T	52,61

8523	31.5.1977	3ft 0in	0-6-0DMF	100hp	Gardner 6LW	15T [HC DM1427]	102
8578	23.10.1978	2ft 1½in	0-6-0DMF	100hp	Gardner 6LW	15T-8 cwt	3
8579	31.1.1979	2ft 1in	0-6-0DMF	100hp	Gardner 6LW	15T-8 cwt	3
8831	25.10.1978	2ft 3in	4wDHF	28hp	Perkins 3152	4T	71
8832	1.11.1978	2ft 3in	4wDHF	28hp	Perkins 3152	4T	71
8842	28.3.1980	2ft 0in	0-6-0DMF	100hp	Gardner 6LW	15T [HC DM1442]	73
8843	18.6.1980	2ft 0in	0-6-0DMF	100hp	Gardner 6LW	15T [HC DM1443]	73
8844	10.10.1980	2ft 0in	0-6-0DMF	100hp	Gardner 6LW	15T [HC DM1444]	
							42,73
8901	24.7.1981	4ft 8½in	0-6-0DM	204hp	Rebuild of HC D1189/60	30T	14
8951	27.2.1981	2ft 6in	4wDHF#	91hp	Perkins 6354	11.4T	52
8952	30.4.1981	3ft 0in	4wDHF#	91hp	Perkins 6354	11.4T	78
8954	30.3.1979	2ft 6in	4wDHF#	91hp	Perkins 6354	11.4T	52
8955	29.6.1981	3ft 0in	4wDHF#	91hp	Perkins 6354	11.4T	78
8956	30.7.1981	3ft 0in	4wDHF#	91hp	Perkins 6354	11.4T	78
8957	30.7.1981	3ft 0in	4wDHF#	91hp	Perkins 6354	11.4T	78

Rack fitted locomotives

R. & W. HAWTHORN, LESLIE & CO LTD, Newcastle-on-Tyne · HL

2454	3.1900	4ft 8½in	0-4-0ST	OC	14 x 20	3ft6in	32,39,74,81
2464	16.6.1900	4ft 8½in	0-4-0ST	OC	14 x 20	3ft6in	1,18,32,76
2490	5.5.1901	4ft 8½in	0-4-0ST	OC	14 x 20	3ft6in	74
2559	23.7.1903	4ft 8½in	0-4-0ST	OC	14 x 20	3ft6in	84
2879	29.7.1911	4ft 8½in	0-6-2T	OC	14 x 22	3ft6in	69
2971	21.11.1912	4ft 8½in	0-4-0ST	OC	16 x 22	3ft6in	84
2980	20.2.1913	4ft 8½in	0-6-0T	OC	16 x 22	3ft10in	76
3002	19.9.1913	4ft 8½in	0-6-0ST	IC	15 x 22	3ft6½in	39
3197	26.7.1916	4ft 8½in	0-6-0ST	OC	14 x 22	3ft6in	40
3307	2.3.1918	4ft 8½in	0-4-0ST	OC	14 x 22	3ft6in	24
3480	5.2.1921	4ft 8½in	0-4-0ST	OC	14 x 22	3ft6in	49
3658	22.7.1926	4ft 8½in	0-6-0ST	IC	17 x 26	4ft1in	18,32
3676	4.10.1927	4ft 8½in	0-6-0T	OC	16 x 24	3ft10in	4,56,76,90
3726	17.8.1928	4ft 8½in	0-6-0ST	IC	16 x 24	4ft1in	16,69
3771	4.11.1930	4ft 8½in	0-4-0ST	OC	14 x 22	3ft6in	29,51,69
3899	17.12.1936	4ft 8½in	0-6-0ST	OC	16 x 24	4ft1in	46
3910	1.6.1937	4ft 8½in	0-4-0ST	OC	14 x 22	3ft6in	29,51

HORWICH WORKS, LYR/LMSR/BR · Hor

[08679]	8.1959	4ft 8½in	0-6-0DE	350hp	EE 6KT	48T	67

JOHN FOWLER & CO (LEEDS) LTD, Hunslet, Leeds · JF

22287	28.21938	4ft 8½in	0-4-0DM	40hp	Fowler 4B	10T	7,11,36,79
22558	26.5.1939	4ft 8½in	0-4-0DM	80hp	Fowler 6A	20T	6,27,36
							82,97
22881	3.11.1939	4ft 8½in	0-4-0DM	150hp	Fowler 4C	29T	95
22887	29.12.1939	4ft 8½in	0-4-0DM	150hp	Fowler 4C	26T	88

KITSON & CO LTD, AIREDALE FOUNDRY, Leeds — K

3819	4.5.1899	4ft 8½in	0-6-0ST	IC	17½x26	4ft6½in		40
3881	14.1.1899	4ft 8½in	0-4-0ST	OC	14 x 21	3ft2½in		11,64
5182	4.4.1919	4ft 8½in	0-6-0T	IC	17½x26	4ft6in		37

KERR, STUART & CO LTD, CALIFORNIA WORKS, Stoke-on-Trent — KS

3075	14.9.1917	4ft 8½in	0-6-0T	OC	17 x 24	4ft0in	50T	70
4080	27.8.1919	4ft 8½in	0-6-0T	OC	15 x 20	3ft9in	34T	69

LOWCA ENGINEERING CO LTD, Lowca, Whitehaven, Cumberland — LE

242	1907	4ft 8½in	0-4-0ST	OC	14 x 20		29

MARKHAM & CO LTD, Chesterfield, Derbyshire — Mkm

	1909	4ft 8½in	0-4-0ST	OC	14 x 21	3ft6in	17,61
	1914	4ft 8½in	0-4-0ST	OC	14 x 21	3ft6in	65

MOTOR RAIL LTD, SIMPLEX WORKS, Bedford — MR

7218	5.9.1938	2ft 0in	4wDM	20/28hp	Dorman 2DWD	2½T	33,66
7406	2.10.1939	2ft 00.in	4wDM	20/28hp	Dorman 2DWD	2½T	19,21
7606	18.4.1939	1ft 11in	4wDM	16/24hp	Ailsa Craig	2½T	21,28
8814	17.31943	3ft 0in	4wDM	20/28hp	Dorman 2DWD	2½T	19,21,28
9695	17.4.1952	2ft 2in	4wDM	20/28hp	Dorman 2DWD	3½T	19,60,91
9696	24.6.1952	3ft 0in	4wDM	20/28hp	Dorman 2DWD	3½T	19,49, 60
9697	30.6.1952	2ft 0in	4wDM	20/28hp	Dorman 2DWD	3½T	60,66
40S280	30.11.1966	3ft 0in	4wDM	40hp	Dorman 2LB	4½T	20,50,66,77

Unidentified locos by MR are listed on pages 60,68 & 77

METROPOLITAN-VICKERS ELECTRICAL CO LTD, Manchester — MV

880	6.8.1953	2ft 0in	4wBEF	90hp	DBF12 [Bg 3400]	8
881	26.10.1953	2ft 0in	4wBEF	90hp	DBF12 [Bg 3401]	8
981	1958	2ft 6in	4wBEF	90hp	DBF12	52
982	1958	2ft 6in	4wBEF	90hp	DBF12	52
983	1958	2ft 6in	4wBEF	90hp	DBF12	52
984	1958	2ft 6in	4wBEF	90hp	DBF12	52
985	1958	2ft 6in	4wBEF	90hp	DBF12	52
986	1958	2ft 6in	4wBEF	90hp	DBF12	52
987	1958	2ft 6in	4wBEF	90hp	DBF12	52
1180	29.9.1960	2ft 6in	4wBEF	90hp	DBF12 [Bg 3563]	52

MANNING WARDLE & CO LTD, BOYNE ENGINE WORKS, Leeds   MW

541	1.4.1875	4ft 8½in	0-6-0ST	IC	13 x 18	3ft0in	M	12
1207	30.12.1890	4ft 8½in	0-6-0ST	IC	15 x 22	3ft6in	Spl	94
1379	1.7.1898	4ft 8½in	0-6-0ST	IC	13 x 18	3ft0in	M	26,89
1393	10.11.1898	4ft 8½in	0-6-0ST	IC	13 x 18	3ft0in	M	2
1589	12.12.1902	4ft 8½in	0-6-0ST	IC	15 x 22	3ft9in	O	46
1667	17.6.1906	4ft 8½in	0-6-0ST	IC	12 x 17	3ft0in	K	2,43,100
1690	2.10.1906	4ft 8½in	0-6-0ST	IC	12 x 18	3ft0in	L	100
1842	14.1.1915	4ft 8½in	0-4-0ST	OC	14 x 20	3ft1½in	14in spl	49
1843	19.4.1915	4ft 8½in	0-4-0ST	OC	14 x 20	3ft1½in	14in spl	39,56
1902	24.5.1916	4ft 8½in	0-4-0ST	OC	14 x 20	3ft1½in	14in spl	17
1968	24.10.1918	4ft 8½in	0-4-0ST	OC	14 x 20	3ft1½in	14in spl	1,31

NOTE that 'alt' = altered and 'Spl' = Special.

NEWTON CHAMBERS & CO LTD, Chapeltown, Yorkshire   NC

	c1939	4ft 8½in	4wPM	36

NORTH BRITISH LOCOMOTIVE CO LTD, HYDE PARK, Glasgow   NBH

21520	1917	4ft 8½in	0-6-0T	IC	17 x 22	4ft2in	40

PECKETT & SONS LTD, ATLAS WORKS, Bristol   P

497	27.1.1891	4ft 8½in	0-6-0ST	IC	16 x 22	3ft10in	X	64
701	12.9.1898	4ft 8½in	0-4-0ST	OC	14 x 20	3ft2in	W4	1,31,74,75,76
824	9.1.1900	4ft 8½in	0-6-0ST	IC	16 x 22	3ft10in	X	18
836	23.7.1900	4ft 8½in	0-6-0ST	IC	16 x 22	3ft10in	X	18
951	3.11.1903	4ft 8½in	0-6-0ST	IC	16 x 22	3ft10in	X	99
992	2.6.1905	4ft 8½in	0-4-0ST	OC	14 x 20	3ft2in	W4	25,63
1114	12.6.1907	4ft 8½in	0-4-0ST	OC	14 x 20	3ft2½in	W5	1,56
1117	4.12.1907	4ft 8½in	0-6-0ST	OC	14 x 20	3ft7in	B2	13
1219	20.6.1910	4ft 8½in	0-6-0ST	IC	16 x 22	3ft10in	X2	22,84
1242	26.10.1911	4ft 8½in	0-6-0ST	IC	16 x 22	3ft10in	X2	4,55
1303	7.4.1913	4ft 8½in	0-6-0ST	IC	16 x 22	3ft10in	X2	46,76
1454	3.1.1917	4ft 8½in	0-4-0ST	OC	14 x 20	3ft2½in	W5	16,39,51,88
1518	24.7.1919	4ft 8½in	0-6-0ST	IC	16 x 22	3ft10in	X2	6,7,31
1519	17.11.1919	4ft 8½in	0-6-0ST	IC	16 x 22	3ft10in	X2	27,46
1578	11.7.1921	4ft 8½in	0-6-0ST	IC	16 x 22	3ft10in	X2	55
1627	21.2.1924	4ft 8½in	0-4-0ST	OC	14 x 20	3ft2½in	W5	7,79
1634	28.2.1927	4ft 8½in	0-6-0ST	IC	16 x 22	3ft10in	X2	16
1650	22.7.1924	4ft 8½in	0-4-0ST	OC	14 x 20	3ft2½in	W5	24
1653	7.1.1925	4ft 8½in	0-4-0ST	OC	14 x 20	3ft2½in	W5	16
1765	17.12.1928	4ft 8½in	0-6-0ST	IC	16 x 22	3ft10in	X2	16
1801	2.2.1931	4ft 8½in	0-6-0ST	OC	14 x 22	3ft7in	B2	98,99
1891	5.2.1940	4ft 8½in	0-6-0ST	IC	16 x 22	3ft10in	X2	24,32,55
2026	22.6.1942	4ft 8½in	0-4-0ST	OC	14 x 22	3ft2½in	W7	56,90
2108	2.1.1950	4ft 8½in	0-4-0ST	OC	15 x 23	3ft7in	E1	25,46,64

PEARSON & KNOWLES COAL & IRON CO LTD, Wigan   P&K

1917	4ft 8½in	0-6-0ST	IC		98

RUSTON & HORNSBY LTD, Lincoln   RH

189958	14.3.1938	2ft 0in	4wDM	16/20HP	Ruston 2VSO	2¾T	41,61
192843	6.9.1938	2ft 0in	4wDM	16/20HP	Ruston 2VSO	2¾T	33,92,96
192848	5.9.1938	2ft 0in	4wDM	16/20HP	Ruston 2VSO	2¾T	96
198261	13.11.1939	2ft 0in	4wDM	11/13HP	Ruston 2VTO	3T	98
211644	10.1.1942	2ft 0in	4wDM	20DL	Ruston 2VSO	2¾T	15,17
211650	19.2.1942	2ft 0in	4wDM	20DL	Ruston 2VSO	2¾T	68
221590	22.4.1943	2ft 0in	4wDM	20DL	Ruston 2VSO	2¾T	41,84
222066	26.8.1943	2ft 0in	4wDM	20DL	Ruston 2VSO	2¾T	61,72
223699	31.1.1944	2ft 0in	4wDM	20DL	Ruston 2VSO	2¾T	3
223742	25.3.1944	2ft 0in	4wDM	20DL	Ruston 2VSO	2¾T	98
223748	25.3.1944	2ft 0in	4wDM	20DL	Ruston 2VSO	2¾T	17
249550	27.11.1947	3ft 0in	4wDMF	48DLG	Ruston 4VRH	7T	14
249552	28.11.1947	3ft 0in	4wDMF	48DLG	Ruston 4VRH	7T	14
249554	19.12.1947	3ft 0in	4wDMF	48DLG	Ruston 4VRH	7T	14,101
249557	14.11.1947	2ft 0in	4wDMF	48DLG	Ruston 4VRH	7T	41,62,84
249559	14.11.1947	2ft 0in	4wDMF	48DLG	Ruston 4VRH	7T	41,62,84,102
249561	28.11.1947	2ft 0in	4wDMF	48DLG	Ruston 4VRH	7T	62,84
249563	27.11.1947	2ft 0in	4wDMF	48DLG	Ruston 4VRH	7T	25,62
249565	3.12.1947	2ft 2in	4wDMF	48DLG	Ruston 4VRH	7T	85,102
249567	12.12.1947	2ft3in	4wDMF	48DLG	Ruston 4VRH	7T	1,5,10,33,92
252863	5.11.1947	2ft 0in	4wDM	20DL	Ruston 2VSO	2¾T	72
252864	6.11.1947	2ft 0in	4wDM	20DL	Ruston 2VSO	2¾T	72
256275	28.4.1948	1ft 11in	4wDMF	48DL	Ruston 4VRO	7T	1,33,92
268857	5.11.1948	2ft 0in	4wDMF	48DLZ	Ruston 4VRH	7T	3,84,92
268860	29.4.1949	1ft 11in	4wDMF	48DLZ	Ruston 4VRH	7T	1,21,28,34
268868	26.10.1950	2ft 2in	4wDMF	48DLZ	Ruston 4VRH	7T	1,5
268870	15.8.1950	2ft 0in	4wDM	48DLZ	Ruston 4VRH	7T	85
331263	24.10.1952	2ft 0in	4wDM	20DL	Ruston 2VSH	3¼T	72
347748	15.5.1958	4ft 8½in	0-6-0DM	LWS	Paxman 8RPH	44T	4,56
347749	2.6.1958	4ft 8½in	0-6-0DM	LWS	Paxman 8RPH	44T	4,56
370543	12.7.1954	2ft 0in	0-4-0DMF	LHG	Ruston 4YE	10T	85
382808	1.3.1955	2ft 0in	4wDMF	40DL	Ruston 3VRH	4½T	49,91
384146	13.2.1956	4ft 8½in	0-6-0DE	165DE	Ruston 6VPH	30T	2
441424	12.11.1959	2ft 0in	4wDMF	48DLZ	Ruston 4VRH	7T	95
476139	19.4.1963	4ft 8½in	4wDH	88DS	Ruston 4VPH	17T	81
481552	28.9.1962	2ft 0in	4wDMF	48DLG	Ruston 4VRH	7T	85
480679	10.11.1961	2ft 6in	4wDMF	20DLG	Ruston 2VSH	3½T	95

NOTE that the details in column 5 are the RH Classes, which indicate the horsepower. Details for the later 'lettered' classes are: LWS - 333hp

ROLLS ROYCE LTD, SENTINEL WORKS, Shrewsbury   RR

10203	5.1.1965	4ft 8½in	0-4-0DH	325hp	R-R C8SFL	31T	25,81
10220	29.6.1965	4ft 8½in	0-6-0DH	325hp	R-R C8SFL	48T	81
10223	23.3.1965	4ft 8½in	0-6-0DH	325hp	R-R C8SFL	48T	18,56
10261	14.12.1966	4ft 8½in	0-6-0DH	325hp	R-R C8SFL	48T	56

ROBERT STEPHENSON & CO, FORTH BANKS WORKS,
Newcastle on Tyne   RS

2893	18.6.1898	4ft 8½in	0-6-0ST	IC	16 x 24	4ft0½in		36,37
3094	1902	4ft 8½in	0-6-0ST	OC	16 x 20	3ft6in		67

ROBERT STEPHENSON & HAWTHORNS LTD   RSH
D   **DARLINGTON WORKS, Co. Durham**   RSHD
N   **FORTH BANKS WORKS, Newcastle-upon-Tyne**   RSH

N	6942	26.4.1938	4ft 8½in	0-6-0ST	IC	17 x 26	4ft1in		18,32
N	7086	29.4.1943	4ft 8½in	0-6-0ST	IC	18 x 26	4ft3in	48T	2,100
N	7094	15.7.1943	4ft 8½in	0-6-0ST	IC	18 x 26	4ft3in	48T	70
N	7096	8.8.1943	4ft 8½in	0-6-0ST	IC	18 x 26	4ft3in	48T	80
N	7100	4.9.1943	4ft 8½in	0-6-0ST	IC	18 x 26	4ft3in	48T	80,92,95
N	7161	19.9.1944	4ft 8½in	0-6-0ST	IC	18 x 26	4ft3in	48T	80
N	7162	20.9.1944	4ft 8½in	0-6-0ST	IC	18 x 26	4ft3in	48T	80
N	7164	4.10.1944	4ft 8½in	0-6-0ST	IC	18 x 26	4ft3in	48T	80,92,95
N	7204	13.3.1945	4ft 8½in	0-6-0ST	IC	18 x 26	4ft3in	48T	18
N	7804	15.11.1954	4ft 8½in	0-4-0WE	80hp				82
N	7847	14.9.1955	4ft 8½in	0-6-0F	OC	20 x 18	3ft0½in		53,54
N	7867	7.1.1957	4ft 8½in	0-6-0DM	204hp	[DC 2580]			51,56
D	7879	23.9.1957	4ft 8½in	0-6-0DM	204hp	[DC 2602]			94
D	7918	7.1959	4ft 8½in	0-6-0DM	204hp	[DC 2620]			51
D	8102	3.1960	4ft 8½in	0-6-0DM	204hp	[DC 2661]			37,67
D	8176	2.1961	4ft 8½in	0-6-0DM	204hp	[DC 2698]			24,56
D	8181	3.1961	4ft 8½in	0-6-0DM	204hp	[DC 2703]			49,65,70,87
D	8185	5.1961	4ft 8½in	0-6-0DM	204hp	[DC 2707]			29,56,65
D	8186	5.1961	4ft 8½in	0-6-0DM	204hp	[DC 2708]			29,32
D	8187	5.1961	4ft 8½in	0-6-0DM	204hp	[DC 2709]			24,29,49
D	8191	6.1961	4ft 8½in	0-6-0DM	204hp	[DC 2713]			19,29,56,86
D	8193	7.1961	4ft 8½in	0-6-0DM	204hp	[DC 2715]			29,51,56,86
D	8194	8.1961	4ft 8½in	0-6-0DM	204hp	[DC 2716]			51,56
D	8195	8.1961	4ft 8½in	0-6-0DM	204hp	[DC 2717]			56
D	8196	8.1961	4ft 8½in	0-6-0DM	204hp	[DC 2718]			4,56

NOTE that 204hp locos were 29¾T with Gardner 8L3 engines

SENTINEL (SHREWSBURY) LTD, Battlefield, Shrewsbury   S

9394	30.8.1950	4ft 8½in	4wVBT	VCG	6¾ x 9	2ft6in	100hp	25T	6,7 64,79
9395	30.8.1950	4ft 8½in	4wVBT	VCG	6¾ x 9	2ft6in	100hp	25T	49
9398	1.1.1950	4ft 8½in	4wVBT	VCG	6¾ x 9	2ft6in	100hp	25T	6,65
9400	14.7.1950	4ft 8½in	4wVBT	VCG	6¾ x 9	2ft6in	100hp	25T	6,82,100
9401	14.7.1950	4ft 8½in	4wVBT	VCG	6¾ x 9	2ft6in	100hp	25T	11,82 97,100
9526	26.1.1952	4ft 8½in	4wVBT	VCG	6¾ x 9	2ft6in	100hp	24T	29
9548	7.11.1952	4ft 8½in	4wVBT	VCG	6¾ x 9	3ft2in	200hp	34T	70,87
9552	21.1.1953	4ft 8½in	4wVBT	VCG	6¾ x 9	3ft2in	200hp	34T	4,48 56,90
9557	23.4.1953	4ft 8½in	4wVBT	VCG	6¾ x 9	2ft6in	100hp	24T	12,79,97
9570	23.2.1954	4ft 8½in	4wVBT	VCG	6¾ x 9	3ft2in	200hp	34T	7,11,82

9616	12.12.1957	4ft 8½in	4wVBT	VCG 6¾ x 9	3ft2in	200hp 34T	7,12,82
9629	25.3.1957	4ft 8½in	4wVBT	VCG 6¾ x 9	2ft6in	100hp 24T	87
10003	7.5.1959	4ft 8½in	4wDH	230hp	R-R C6SFL	34T	65
10058	16.3.1961	4ft 8½in	4wDH	230hp	R-R C6SFL	34T	31
10059	16.3.1961	4ft 8½in	4wDH	230hp	R-R C6SFL	34T	94
10176	14.2.1964	4ft 8½in	4wDH	230hp	R-R C6SFL	32T	7
10180	27.2.1964	4ft 8½in	0-6-0DH	311hp	R-R C8SFL	48T	18,51
10181	20.3.1964	4ft 8½in	0-6-0DH	311hp	R-R C8SFL	48T	57,76,87

SWINDON WORKS, BR   Sdn

[BR D2182]	3.1962	4ft 8½in	0-6-0DM	204hp	Gardner 8L3	30T	94
[BR D2199]	6.1961	4ft 8½in	0-6-0DM	204hp	Gardner 8L3	30T	8,46,71
[BR D2373]	8.1961	4ft 8½in	0-6-0DM	204hp	Gardner 8L3	30T	4,56,90

THOMAS HILL (ROTHERHAM) LTD, Kilnhurst, S Yorkshire   TH

142C	12.8.1964	4ft 8½in	4wDH	311hp	R-R C8SFL	36T	67,79,82
156C	9.9.1965	4ft 8½in	4wDH	311hp	R-R C8SFL	36T	7,82
158C	20.10.1965	4ft 8½in	4wDH	178hp	R-R C6NFL	25T	7,31,100
171C	6.10.1966	4ft 8½in	4wDH	173hp	R-R SF65C	25T	25
250V	11.4.1974	4ft 8½in	0-6-0DH	370hp	R-R C8TFL	48T	57
SE102	.1976	2ft 4in	4wBERF			5T	48

UNDERGROUND MINING MACHINERY LTD, Aycliffe, Co. Durham   UMM

23.07	1970	500mm	2adDHF	23hp	Coventry Victor Vixen	63
40.002	1971	500mm	2adDHF	40hp	Perkins 4.203	63

HUGO AECKERLE & CO, Hamburg, Germany   Ulk

2005	pre 1972	4ft 8½in	4wDH	R/R	94

VULCAN FOUNDRY LTD, Newton le Willows, Lancs   VF

878	1880	4ft 8½in	0-6-0ST	IC 17½x26	4ft7½in		98
5295	1945	4ft 8½in	0-6-0ST	IC 18 x 26	4ft3in	48T	37
5296	1945	4ft 8½in	0-6-0ST	IC 18 x 26	4ft3in	48T	37
D209	2.1953	4ft 8½in	0-6-0DM	204hp	[DC 2483]		19,24,56,76
D210	3.1953	4ft 8½in	0-6-0DM	204hp	[DC 2484]		49,56
D257	10.1954	4ft 8½in	0-6-0DM	204hp	[DC 2529]		56
D268	8.1955	4ft 8½in	0-6-0DM	204hp	[DC 2542]		7
D274	10.1955	4ft 8½in	0-6-0DM	204hp	[DC 2548]		4,56,90
D278	11.1955	4ft 8½in	0-6-0DM	204hp	[DC 2552]		16,70
D288	5.1956	4ft 8½in	0-6-0DM	204hp	[DC 2562]		54
D289	5.1956	4ft 8½in	0-6-0DM	204hp	[DC 2563]		31

NOTE that 204hp locos were 29¾T with Gardner 8L3 engines

W. G. BAGNALL LTD, CASTLE ENGINE WORKS, Stafford   WB

2105	5.11.1919	4ft 8½in	0-4-0ST	OC	14 x 20	3ft6½in	24¾T	7
2223	3.10.1924	4ft 8½in	0-6-0ST	OC	15 x 20	3ft7in	30½T	3,47,56
2752	9.9.1944	4ft 8½in	0-6-0ST	IC	18 x 26	4ft3in	48T	2
2761	30.11.1944	4ft 8½in	0-6-0ST	IC	18 x 26	4ft3in	48T	80

WELLMAN, SMITH OWEN ENGINEERING CORPORATION LTD,
Darlaston, Staffordshire   WSO

	4ft 8½in	0-4-0RE				74

YORKSHIRE ENGINE CO LTD, MEADOW HALL WORKS, Sheffield   YE

118	1869	4ft 8½in	0-4-0ST	OC	10 x 16	3ft0in		65,74
119	1869	4ft 8½in	0-4-0ST	OC	10 x 16	3ft0in		1,34,65
120	1869	4ft 8½in	0-4-0ST	OC	10 x 16	3ft0in		34,65
478	1892	4ft 8½in	0-4-0ST	OC	14 x 20	3ft3in		1,32,34,65,74
479	$ 1.1891	4ft 8½in	0-4-0ST	OC	14 x 20	3ft3in		79,97
483	11.1895	4ft 8½in	0-4-0ST	OC	14 x 20	3ft3in		97
610	3.1900	4ft 8½in	0-4-0ST	OC	14 x 20	3ft3in		97
832	5.1905	4ft 8½in	0-4-0ST	OC	15 x 20	3ft6in		63
1021	# 8.11.1909	4ft 8½in	0-6-0ST	OC	14 x 20	3ft3in	Class G	51,70,87
1026	11.1910	4ft 8½in	0-4-0ST	OC	14 x 20	3ft3in	Class G	26,92,95
1027	1.1912	4ft 8½in	0-4-0ST	OC	14 x 20	3ft3in	Class G	26,92
1787	4.1922	4ft 8½in	0-6-0ST	OC	14 x 20	3ft3in	Class G	14,17,22 40,84
1823	1.1923	4ft 8½in	0-6-0ST	OC	14 x 20	3ft3in	Class G	56,90
1889	9.1923	4ft 8½in	0-6-0ST	OC	14 x 20	3ft3in	Class G	6,11,83
1891	4.1924	4ft 8½in	0-4-0ST	OC	14 x 20	3ft3in	Class G	11,32,81
2240	7.1929	4ft 8½in	0-6-0ST	OC	15 x 22	3ft4in		51,86,87
2241	20.6.1929	4ft 8½in	0-6-0ST	OC	15 x 22	3ft4in		86
2305	4.1931	4ft 8½in	0-6-0ST	OC	15 x 22	3ft5in		90
2473	5.9.1949	4ft 8½in	0-4-0ST	OC	16 x 22	3ft6in		7,12,37,64
2474	23.9.1949	4ft 8½in	0-4-0ST	OC	16 x 22	3ft6in		6,12,64,99
2485	30.3.1951	4ft 8½in	0-6-0ST	OC	16 x 24	3ft8in		51
2521	6.5.1953	4ft 8½in	0-6-0ST	OC	16 x 24	3ft8in	Type 1	94
2569	11.5.1955	4ft 8½in	0-6-0ST	IC	18 x 26	4ft3in	48T	67
2729	3.12.1958	4ft 8½in	0-4-0DE	200hp	R-R C6SFL	30T		7,81
2913	23.2.1965	4ft 8½in	0-6-0DH	375hp	Cummins NT400	51T		16,57,87
2939	26.3.1965	4ft 8½in	0-6-0DH	375hp	Cummins NT400	51T		57,70,87

$ dated 1892 on works plate.
# alternative YE sources state 10.1909.

NINE ELMS WORKS, London & South Western Railway   9E

[LSWR 92]	12.1892	4ft 8½in	0-4-0T	OC	16 x 22	3ft9in	33T	26,80

1.   RICHARD (MW 1968 of 1919) at Aldwarke Main Colliery
      on 16th April 1949.   (G. Alliez)

2.   VICTORY (AE 1834 of 1919) at Askern Main Colliery
      in April 1959   (Frank Jones)

3.   HARRY No.73 (HE 6661 of 1966) at Barnburgh Main Colliery
        on 30th March 1978   (Adrian Booth)

4.   No.1 (P 1518 of 1919) at Barrow Colliery
        on 5th July 1953   (C.H.A. Townley, IRS Collection)

5.   MR 8814 of 1943 at Cadeby Colliery
        on 10th October 1965

                                                    (Roger Hateley)

6.   No.26 (AB 1498 of 1918) at Cortonwood Colliery
        on 12th October 1965

                                                    (A.R. Ethernigton)

7.   DARFIELD No.2 (HE 3805 of 1953) at Darfield Colliery
on 11th June 1968   (Adrian Booth)

8.   HE 7273 of 1973 at Darfield Colliery
on 27th March 1978   (Adrian Booth)

9. DINNINGTON No.3 (S 9526 of 1951) at Dinnington Colliery
    shortly after delivery   (IRS, collection of Brian Webb)

10. BILL (AE 1920 of 1924) at Dinnington Colliery
    on 30th May 1968   (Adrian Booth)

11.   No.52 (HE 4503 of 1955) and No.28 (HE 5514 of 1959) at Dinnington Colliery
         on 20th April 1979   (Adrian Booth)

12.   HATFIELD 4 (AE 1448 of 1902) at Hatfield Colliery
         on 26th May 1964   (K.J. Cooper, IRS collection)

13.   GB 1348 of 1934 at Handsworth Nunnery (High Hazels) Coking Plant
on 11th May 1963   (A.R. Etherington)

14.   VICTORY (AB 1654 of 1920) at Handsworth (High Hazels) Colliery
on 23rd March 1954   (Bernard Mettam)

15.   HC D1342 of 1966 at Hickleton Colliery on 27th August 1979
(Adrian Booth)

16.   P 2108 of 1950 and No.5 (P 1303 of 1913) at Houghton Main Colliery
on 27th May 1962   (Roger Hateley)

17.   MANVERS MAIN 42 (HC 1690 of 1937) at Manvers Main Colliery
        on 16th September 1961   (J.A. Peden, IRS Collection)

18.   CARL No.61 (HC D1154 of 1959) at Manvers Main Colliery
        on 6th February 1969   (Adrian Booth)

19.   11 (YE 1823 of 1922) at Manvers Main Colliery
       on 12th October 1968   (J.A. Peden, IRS Collection)

20.   No.44 WILF (P 1891 of 1940) at Manvers Main Colliery
       on 10th October 1965   (Brian Webb, IRS Collection)

21.   Ex BR D2238 TOM (/VF D288/DC 2562 of 1955) at Manvers Main Colliery
on 7th June 1969   (Adrian Booth)

22.   CHARLES (Mkm of 1909) at Markham Main Colliery
on 26th February 1963   (Roger Hateley)

23.   P 992 of 1905 at Mitchell Main Colliery
          on 9th September 1951   (K.J. Cooper, IRS Collection)

24.   HC 285 of 1889 at New Stubbin Colliery
          in about August 1950   (Frank Jones)

25.   YE 120 of 1869 at New Stubbin Colliery
shortly before scrapping in 1954   (Brian Rumary Collection)

26.   No.55 (HE 6059 of 1962) at New Stubbin Colliery
on 19th April 1978   (Adrian Booth)

27.   Ex BR D2322 (RSHD 8181/DC 2703 of 1961) at Orgreave Colliery
       on 13th April 1979   (Adrian Booth)

28. No.6 (Bg 3490/EE 2392 of 1958) at Silverwood Colliery on 11th June 1978
                                                            (Adrian Booth)

29.   TL13 (TH 142C of 1964) at Skiers Spring Colliery
      on 25th August 1975   (Adrian Booth)

30.   GB 2543 of 1955 at Smithywood Coking Plant
      on 25th May 1979   (Adrian Booth)

31.   No.5 (YE 610 of 1900) at Wharncliffe Silkstone Colliery
        on 2nd June 1956   (Bernard Mettam)

32.   No.5 FORWARD (MW 1690 of 1906) at Yorkshire Main Colliery
        on 6th September 1958   (J.A. Peden, IRS Collection)